T0379797

Math Intervention

Grades 3–5

Help all of your students reach success in math! This essential book, from bestselling author and consultant Jennifer Taylor-Cox, is filled with suggestions that teachers and RTI/MTSS specialists can use to target instruction for struggling students in grades 3–5. You'll learn how to diagnose academic weaknesses, differentiate instruction, use formative assessments, offer corrective feedback, and motivate students with games and activities.

The book's practical features include . . .

- Directions for incorporating formative assessments;
- Explanations of successful strategies for intervention;
- Important math terms to use with students;
- Games for active learning with printable boards;
- Cognitive demand questions ranging from easy to complex; and
- Rigorous problems to help you gather pre and post data.

In this enhanced second edition, you'll find correlations to the Common Core throughout, as well as a variety of brand new, rigorous problems designed to mirror those on CCSS assessments.

Bonus! The book is accompanied by free eResources on our website, www.routledge.com/9781138915695. These eResources include an Answer Key with Scoring Guide and a handy Progress Monitoring Tool that you can use to track each student's growth, record notes, and share data with parents, administrators, and other educators. The eResources also contain printable versions of the games in the book so that you can easily download and print them for classroom use.

Jennifer Taylor-Cox is the owner of Taylor-Cox Instruction, LLC. She serves as an educational consultant for numerous districts across the United States and is the author of seven Routledge Eye On Education books.

Also Available from Routledge Eye On Education
(www.routledge.com/eyeoneducation)

**Math Intervention, Grades PreK–2, 2nd Edition:
Building Number Power with Formative Assessments,
Differentiation, and Games**
Jennifer Taylor-Cox

**Family Math Night, K–5, 2nd Edition:
Common Core State Standards in Action**
Jennifer Taylor-Cox

**Family Math Night:
Middle School Math Standards in Action**
Jennifer Taylor-Cox and Christine Oberdorf

**Solving Behavior Problems in Math Class:
Academic, Learning, Social, and Emotional Empowerment,
Grades K–12**
Jennifer Taylor-Cox

**Using Formative Assessment to Drive Mathematics
Instruction in Grades PreK–2**
Jennifer Taylor-Cox and Christine Oberdorf

**Using Formative Assessment to Drive Mathematics
Instruction in Grades 3–5**
Jennifer Taylor-Cox and Christine Oberdorf

**Motivating Struggling Learners:
10 Ways to Build Student Success**
Barbara R. Blackburn

Rigor Is Not a Four-Letter Word, 2nd Edition
Barbara R. Blackburn

**Less Is More in Elementary School:
Strategies for Thriving in a High-Stakes Environment**
Renee Rubin, Michelle Abrego, and John Sutterby

**What Schools Don't Teach:
20 Ways to Help Students Excel in School and Life**
Brad Johnson and Julie Sessions

**Making Good Teaching Great:
Everyday Strategies for Teaching with Impact**
Annette Breaux and Todd Whitaker

Math Intervention

Building Number Power

with Formative Assessments, Differentiation, and Games

Grades 3–5

2nd Edition

Jennifer Taylor-Cox

NEW YORK AND LONDON

Second edition published 2016
by Routledge
711 Third Avenue, New York, NY 10017

and by Routledge
2 Park Square, Milton Park, Abingdon, Oxon OX14 4RN

Routledge is an imprint of the Taylor & Francis Group, an informa business

Illustrations by Marida Hines

First edition published by Eye On Education 2009

Library of Congress Cataloging-in-Publication Data
Taylor-Cox, Jennifer.
Math intervention : building number power with formative assessments, differentiation, and games : grades 3–5 / Jennifer Taylor-Cox. — 2nd edition.
pages cm
Includes bibliographical references and index.
ISBN 978-1-138-91568-8 (hardback) — ISBN 978-1-138-91569-5 (pbk.) — ISBN 978-1-315-69011-7 (ebook) 1. Mathematics—Study and teaching (Elementary) 2. Mathematical readiness. 3. Group guidance in education. I. Title. II. Title: Building number power with formative assessments, differentiation, and games : grades 3–5.
QA135.6.T3557 2016
372.7'049—dc23
2015032836

ISBN: 978-1-138-91568-8 (hbk)
ISBN: 978-1-138-91569-5 (pbk)
ISBN: 978-1-315-69011-7 (ebk)

Typeset in Palatino
by Apex CoVantage, LLC

Contents

Acknowledgements

This book is in loving memory of my brother, David Eugene Taylor II, my favorite Superman. Thank you for 46 awesome years.
You will live in my heart forever.

Math Rocks!

Appreciation is extended to reviewers Camyll Bontems and Yisenia González-Reyes, outstanding teachers.

About the Author

Dr. Jennifer Taylor-Cox is an enthusiastic, captivating presenter and well-known educator. She is the owner of **Taylor-Cox Instruction, LLC**. Jennifer serves as a consultant in mathematics education providing professional development opportunities for numerous districts across the United States. Her keynote speeches, workshops, seminars, classroom demonstration lessons, and presentations are always high-energy and insightful.

Jennifer earned her Ph.D. from the University of Maryland and was awarded the "Outstanding Doctoral Research Award" from the University of Maryland and the "Excellence in Teacher Education Award" from Towson University. She served as the president of the Maryland Council of Teachers of Mathematics. Jennifer is the author of many professional articles and books including *Math Intervention, Grades PreK–2, 2nd Edition: Building Number Power with Formative Assessments, Differentiation, and Games; Family Math Night, K–5, 2nd Edition: Common Core State Standards in Action; Family Math Night: Middle School Math Standards in Action; Solving Behavior Problems in Math Class: Academic, Learning, Social, and Emotional Empowerment, Grades K–12; Using Formative Assessment to Drive Mathematics Instruction in Grades PreK–2*; and *Using Formative Assessment to Drive Mathematics* Instruction in Grades 3–5.

Jennifer Taylor-Cox, Ph.D., Educational Consultant
Office: 410-729-5599, Fax: 410-729-3211
Email: Jennifer@Taylor-CoxInstruction.com

Dr. Taylor-Cox knows how to make learning mathematics engaging and fun which motivates learners of all ages! She has a passion for differentiated math instruction. Jennifer believes that all students are entitled to targeted instruction aimed at their specific learning needs. Jennifer lives and has her office in Severna Park, Maryland. She is the mother of three children, and a grandmother. Jennifer truly understands how to connect research and practice in education. Her zeal for mathematics education is alive in her work with educators, students, and parents.

If you are interested in learning more about the professional development opportunities Dr. Taylor-Cox offers, please feel free to contact her at **Taylor-CoxInstruction.com**.

eResources

This book is accompanied by free eResources that you can download and print for classroom use. Go to the book product page at:

www.routledge.com/9781138915695

Click on the tab that says eResources and then click on the links for the eResources you'd like. They will begin downloading to your computer.

List of eResources
Games:

Bonus eResources available as downloads:
Progress Monitoring Tool
Answer Key and Scoring Guide

Introduction

What is Math Intervention?

Math intervention is targeted instruction for students struggling with mathematics. Too often students have difficulty understanding math concepts, and the instruction they receive perpetuates or masks the problems rather than addressing them. Students are left to flail and founder in a confusing sea of procedures and rules. The problems become more profound as the student moves into each new grade. As math educators we can no longer let this happen. We must make math meaningful for students and provide targeted instruction that focuses on the precise academic needs of students. It is essential that we concentrate on repairing students' misconceptions and learning gaps in ways that build their math capacity.

In mathematics there are four interconnected goals: accuracy, efficiency, flexibility, and fluency. Our first priority is to help students with their math accuracy, which means we help them learn how to obtain the correct answer. Next, we address efficiency by helping students learn how to obtain the right answer as quickly as possible as is appropriate for the concept. After correctness and speed, we help students improve flexibility so that they understand how to apply their learning to new situations and in different ways. Flexibility means adaptability. We want students to be able to revise and adjust their thinking in ways that maintain and increase accuracy and efficiency. Finally, we must support students as they build their fluency with mathematics. Fluency relates to the confidence and ease with which students work with mathematics concepts.

Accuracy, efficiency, flexibility, and fluency are goals built upon one another within each math concept. Accuracy comes first because there is no point in having any of the other goals without exactness. We do not want students to increase speed if they are getting the answers wrong. Likewise, there is little need to be flexible and fluent with incorrect answers. Educators should focus on accuracy first, even when it takes a long time for students to come up with the correct answer. As soon as students have established accuracy, we need to move right on to focus on efficiency. Our goal is to help students learn how to maintain accuracy while increasing speed. As soon as the students quickly obtain the right answers, we want to focus on flexibility by providing new situations that require changes in their thinking. The fluency aspect of mathematics can be addressed during any of the three previous goals. We can build students' fluency while increasing accuracy, efficiency, and flexibility.

Think about this scenario. A student is trying to solve the following division problem: There are 67 boxes of cookies. The boxes of cookies are packed in cartons. Each carton holds eight boxes of cookies. How many cartons are needed to pack all the boxes of cookies? The student tackles this problem using a *guess and check* strategy. She starts with 100 cartons and figures out ten cartons hold 800 cookies. Because 800 are way too many, she tries 50 cartons. Then she tries ten cartons followed by nine cartons. She tries eight cartons and realizes there are some leftover boxes of cookies, so she tries nine cartons again. The strategy worked because her answer is correct; therefore, she has accuracy. Yet, the *guess and check* strategy was quite time consuming. If we teach this student to use *repeated subtraction*, her speed will increase. Beginning with 67, she subtracts eight, then eight more, then eight more, and so on until she has three boxes left. She knows these boxes also need a carton and comes up with the correct answer to the problem. She has maintained accuracy and increased her speed. To help this student move to flexibility, we can encourage her to consider *fact families* and *inverse operations*. She knows 8 x 8 = 64 and therefore can apply this knowledge to the division situation. Furthering her flexibility, she compares 64 to 67 and accurately and efficiently finds that nine cartons are needed to hold 67 boxes of cookies. Her confidence and ease have increased based upon the successful experiences.

Timing is very important. Sometimes well-meaning math educators jump too fast to efficiency and flexibility. They impose all of these rules and procedures with intentions of helping students see efficient and flexible ways to solve problems. However, teaching rules and procedures is not the same as teaching concepts. Understanding what to do is not enough. We must teach students why a procedure works so that they better understand the concept. If we only teach rules and procedures, struggling students typically achieve random accuracy (if they remember the rules and procedures) and rarely achieve flexibility or fluency because they do not truly own the knowledge. Math educators need to teach concepts to help students build foundations so that they can understand the math. When concepts are ignored and the focus is solely on rules and procedures, struggling students often develop misconceptions and learning gaps. However, if students understand concepts, it is appropriate to help them increase speed and flexibility by teaching rules and procedures, but we must include reasons and connections to help students make this math meaningful.

Students may return to inefficient strategies when faced with challenging problems. They do so because they want to achieve accuracy. These students benefit from the teacher providing intervention that targets their exact academic needs while challenging them to increase accuracy, efficiency, flexibility, and fluency. While it makes sense to focus first on accuracy then move to efficiency, flexibility, and fluency, the goals are not isolated. We often move back and forth, teaching all four goals at the same time to help students understand the hows and whys of mathematics.

Focus on Number Sense and Computation

Typically, the problems that struggling students have in mathematics relate to number sense and computation. Number sense involves an expansive and inclusive understanding of number and operations that allows a person to make sound judgments and utilize practical and effective math strategies (McIntosh, Reys, & Reys 1993). Number sense is not something that a person has or does not have. Number sense can be taught through a concept-based approach. It is important to examine all of the aspects of number sense to identify which of these concepts our struggling students are missing. General intervention is not the best way to build students' number sense. We must build students' number sense through targeted intervention aimed at the students' exact academic needs.

The Common Core State Standards (2010) require deeper understanding of mathematics skills and concepts. Because these standards are based on rigorous content and application of knowledge, students must build strong foundations in critical math concepts. All of the concepts in this book are interrelated components of number, operations, and base ten. The focus on specific concepts does not imply that we need to teach math concepts in isolation. Productive math instruction is inclusive, interrelated, rigorous, and rich. The reason we need to focus on specific concepts in intervention is because we need to identify gaps in knowledge and misconceptions that students have in their understanding of mathematics. Intervention does not replace math instruction. Intervention is in addition to high quality math instruction.

What Makes Math Intervention Successful?

Students who struggle in mathematics often need explicit instruction. Explicit instruction is provided when teachers give students ongoing feedback, clear models, a variety of examples, time for practice, and opportunities to talk through the hows and whys of the math situation (National Mathematics Advisory Panel 2008). Successful math intervention is precise and focused on students' immediate learning needs.

Diagnosing and addressing students' needs are part of Response- to-Intervention (RTI). Although the use of RTI is common practice in language arts, RTI is completely applicable to mathematics, as well. RTI promotes the identification and diagnosis of potential learning problems by analyzing students' responses to instruction and providing corrective feedback. With RTI students are offered specific interventions and progress is frequently monitored. Tier 2 (small group) and Tier 3 (individual) are excellent group structures for providing the specific interventions presented in this book. Successful mathematics intervention applies the structure of RTI—identify, diagnose, and address students' learning weaknesses early and precisely.

There are several other key components that make math intervention successful. The group structure, quantity and quality of the tasks, use of new strategies, educator's expertise, incorporation of problem solving, and student's level of motivation are some of the factors that influence math intervention success.

Productive math intervention is rarely a whole class endeavor. Math intervention is most effective in small group structures. Students are more likely to reveal gaps in understanding and misconceptions when they are in small groups. Teachers are more likely to find these gaps and misunderstandings when they are working with a small group. Furthermore, the small group structure allows teachers to target the instruction more effectively because they can constantly adjust the instruction. Teachers can tailor the interaction and math discourse to help struggling students find success. Small group instruction provides opportunities for strategic mathematics intervention.

When possible, one-on-one instruction has powerful potential. The teacher can concentrate completely on one student's academic needs. For some of us, one-on-one instruction is only possible in small chunks of time because we have an entire class to teach. For others such as special educators, tutors, paraeducators, assistants, parents, and volunteers, one-on-one instruction is more common. The benefits of one-on-one instruction can be great because this structure provides opportunities for intensive intervention. Students are typically highly engaged and they gain confidence in their abilities.

More is better is not necessarily true when it comes to math intervention. It really depends on what you mean by *more.* A packet of 100 math problems does not make a better mathematics student. More problems do not necessarily help students, especially if students are left alone to solve page after page of problems. In these cases, all the students typically learn is that they are not good at solving math problems—worse yet, that they hate math. To make the *more is better* idea beneficial for struggling students, consider more time, more instruction, more strategies, and more opportunities for success. These are the ingredients needed for successful math intervention.

Drill does kill. It kills the spirit of learning (Taylor-Cox & Kelly 2008). Too much drill is boring, isolated, and unproductive. Struggling students do not need stacks of flash cards. They need motivating math games that provide them with multiple opportunities for learning. Struggling students must have experiences that help them understand the math, not empty, meaningless experiences. They need context and connections so they learn to own the mathematics.

Certainly, struggling students do not need more of the same. Obviously, the original instruction did not work; otherwise they would not be struggling. Struggling students need new strategies and experiences beyond the textbook. We must offer our struggling students targeted instruction aimed at fortifying specific math concepts for students. Struggling students benefit from new and creative ways in which to learn the information.

An educator's expertise plays an important role in successful math intervention for struggling students. One of the recommendations made on the report given by the National Mathematics Advisory Panel involves the expertise of mathematics educators. The panel recommends that "teachers must know in detail the mathematical content they are responsible for teaching and its connections to other important mathematics topics, both prior to and beyond the level they are assigned to teach" (2008 p. 37). Educators must understand the concepts, know which concepts the student needs to work on, and know how to assess the level of understanding the student currently has. Good educators strive to increase their levels of expertise. They gain knowledge through their classroom experiences, pedagogical discourse with other educators, and using a variety of resources. Those providing math intervention can be a veteran or new classroom teacher, special educator, ELL teacher, paraeducator, instructional assistant, teacher aide, volunteer, tutor, and/or parent. The point is for educators to have the expertise and resources necessary to provide successful math intervention for students.

Successful math intervention includes the important processes established within the CCSS for Mathematical Practice (2010). We want our students to persevere, reason, evaluate, use tools strategically, attend to precision, and search for math structure. These essential processes help students apply and adapt critical thinking. The invitation to engage in each of the Mathematical Practices during intervention is present any time, but is particularly active when the teacher asks higher level questions and when students participate in concept-driven games and problem solving situations.

The level of motivation and commitment that a student has influences the effectiveness of math intervention. Games motivate and empower students. Games open up multiple opportunities for meaningful mathematics, and they are fun.

If given a choice, most students would choose to play a game rather than solve math problems on a page in a textbook. Yet the game may actually teach them more about math than they even realize. That's the beauty of games! They are so entertaining that the learning doesn't seem like a chore. This is not to imply that the math will always be easy. Challenges and increased difficulty are important parts of learning mathematics. Math games can be challenging and enjoyable because games serve as a productive way to hook students into successfully engaging in mathematics. The games in this book provide opportunities for students to learn new concepts, practice skills, and apply math thinking to new situations.

How to Use Formative Assessment

Using formative assessment allows us to know what to teach. Formative assessment provides information about what a student currently knows. We can use the formative assessment data to focus on the explicit academic needs of the students.

"Teachers' regular use of formative assessment improves their students' learning, especially if teachers have additional guidance on using the assessment to design and individualize instruction" (National Mathematics Advisory Panel 2008 p. xxiii). Formative assessment comes in a variety of forms. Sometimes it is a performance task. Other times it is a problem or question. Using a formative assessment that is not too time-consuming allows us to devote as much time as possible to the instruction. Interestingly, using formative assessments can actually save time because the information helps the teacher target the instruction. Each of the concepts presented in this book includes a formative assessment that will allow the teacher to uncover and focus on the academic needs of the students.

Monitor Progress

Monitoring students' progress is critically important. As educators we must constantly analyze the progress of students. To accomplish this we need to ask questions:

- Is the student making progress?
- What does the student need to learn next?
- How solid is the student's understanding?
- Does the student need more work with a specific concept?
- Is the student having difficulty maintaining and utilizing specific concepts?
- What misconceptions does the student have?
- Where are the learning gaps?
- Is the student's knowledge incomplete? If so, what is missing?

The answers to these questions help us monitor the progress of students. As we monitor progress, we make decisions about how to target the instruction to meet the exact academic needs of the students.

Struggling students need to understand their progress. As educators monitor the progress of students, they should share it with the students. One way to do so is to regularly meet with students and share evidence of the students' gains in knowledge. The evidence may be formative assessment data and/or the actual preassessments and postassessments. The evidence may be a copy of the game board with notes about what a student did to solve problems or a paper that shows how a student used specific concepts and strategies. Many educators use these and other types of evidence to evaluate a student's performance. The key is to provide constant feedback by sharing this information with the student.

A newly available Progress Monitoring Tool can be downloaded for use. Each math intervention concept is divided into three levels of understanding (I-1, I-2, I-3). The progress Monitoring Tool allows the teacher to capture the degree of understanding, the date of concept focus, and the date of proficiency for each student. It also allows teachers to record student progress on problem solving.

Additional ways to help students monitor their own progress include recording the things they know about the concept, have done with the concept, and goals they have related to the concept *(see page 17 for monitoring tool)*. It is also important for students to evaluate the degree of knowledge of any given concept. Using personal scales to report levels of knowledge is beneficial. To do so, the student places X on a scale (the scale on the monitoring tool is 0–10, but any scale is fine) and dates the X underneath. Later, the student adds other Xs and dates to the scale to show progress. Scales offer a productive way to help students focus on their progress.

A former student, Tony, started the school year struggling in mathematics. He had many misconceptions and learning gaps. We worked on building his understanding of specific math concepts by implementing targeted instruction and playing math games to build conceptual knowledge. Evidence of Tony's progress was archived in an accordion folder. The pre-assessments were stapled to the post- assessments to highlight progress. One day Tony was proudly going through the folder. I took the opportunity to join him and we celebrated his progress. After looking through the papers, Tony announced, "I really learned a lot. I can't believe it all fits in my brain!"

Positive ways to share progress with students also include highlighting what they have learned and noting what they need to work on next. The educator may say to the student, "You really improved your accuracy and speed with doubles. The next concept we will focus on is near doubles." The idea is to celebrate what the student has learned and set the next goals.

Reteach

Sometimes students do not learn what they need to learn during regular classroom instruction or intervention. In these cases, we must focus on reteaching. Effective reteaching is not teaching the exact same thing in the exact same way again and again. If students are not learning, we identify the concept and present it in a different way. It may be necessary to move back to a previous concept to build foundations. Or, we may need to reveal and address a misconception. Reteaching should be targeted to meet the explicit academic needs of the students. If a student does not understand fact families, telling the student, in a louder voice, what fact families are is not effective reteaching! Instead, we must try new ways to help the student visualize and model fact families. It may also be that the student needs more work with addition and subtraction concepts. The point is to make the reteaching match the immediate needs of the student.

How to Use This Book

We need to know and understand each concept so that we can identify and focus on specific instruction. This book is designed to give math educators the tools to provide direct math intervention in the areas of number sense and computation.

Students who struggle in mathematics deserve personalized math intervention. Knowing which number concepts a student is lacking (whether the student is eight years old or twelve years old) is critical. It is not one-size-fits-all math intervention. A fifth grader may be struggling with third grade concepts. The point is not to focus so much on the grade level; instead we should focus on the specific concepts that the student needs to learn.

Each math concept presented in this book includes a brief description to provide the educator with background information aimed at increasing expertise. In essence, all of these concepts are connected. Therefore, many of the games include multiple concepts. However, each concept is highlighted independently to enable the educator to target instruction for students. The most relevant domains of the Common Core State Standards are also included.

A formative assessment is provided for each concept to help the educator find out if the student has gaps in understanding the concept or if the student has a solid understanding of the concept. Sometimes exact quantities and specific numbers are provided. Other times choices are provided. The intent is for the formative assessments to serve as tools for targeted instruction.

Successful strategies are included with each number concept for the purpose of offering helpful hints and directions. A math word box accompanies each concept to provide potential math words that should be modeled by the educator and used by the students. These are not vocabulary words to be taught in isolation. Instead, these are conceptually rich math words that should be used in context.

To further promote math discourse and math thinking, sample questions are included. Each question represents a different *level of cognitive demand.* Using Bloom's (1956) taxonomy for structure, a sample question is listed for recall/ remember, comprehension/understand, application, analysis, evaluation, and synthesis/create. The simplest question (recall) is listed first, followed by increasingly advanced levels of questions through the highest level of cognitive demand needed (synthesis). *Note: Some questions could represent more than one level of cognitive demand.* The educator should decide which level of questioning best suits the student's needs at the time. Students should be encouraged to ask questions, too.

To motivate students and activate learning, a math game that directly connects to the specific concept is offered. During the game, the educator should encourage students to use math words, ask questions, and explain and justify their thinking. Each game lists necessary materials and step-by-step directions. Each game is intended to be a springboard for active learning. The educator may modify instruction based on how the student responds to various components of the game

Each game includes potential content differentiation. If the student has too much difficulty with the concept, the educator is invited to differentiate the content of the game for the student by *moving back* to an easier version of the game. If the student understands the concept, the educator is invited to differentiate the content of the game for the student by *moving ahead* to a more complex version.

Rigorous Problem Solving

A new feature of this revised intervention book is the rigorous problem solving component correlated to each math concept. The problems were developed to mirror the thinking required in CCSS-driven assessments such as the assessments developed by Partnership for Assessment of Readiness for College and Career (PARCC) and Smarter Balanced Assessment Consortium (SBAC). While building familiarity with format can help students improve results, it is the rigor and required critical thinking that are the most important aspects of this new problem solving component. This rigorous problem solving component uses Bloom's (1956) levels of cognitive demand, which direct the type of thinking in conjunction with Webb's (2002) Depth of Knowledge (DOK) levels which target the actual task. While recall and reproduction (DOK level 1) are included, the majority of the problem solving is aimed at reaching Skill and Concept (DOK level 2), Strategic Thinking (DOK level 3), and Extended Reasoning/Thinking (DOK level 4). The problems are also directly linked to the CCSS for Mathematical Practice. Each problem has two versions of the same content at the same level. The two versions can be used to gather pre and post data. Or the first version can be utilized instructionally with the intervention teacher and the second version tackled by each student independently. Student progress can be monitored with the problem solving and the progress monitoring tool.

This math intervention book is a tool for educators. The book includes addition and subtraction through multifaceted number concepts designed for third through fifth grade students. The companion book *Math Intervention: Building Number Power with Formative Assessments, Differentiation, and Games (Grades PreK–2)* includes early number concepts through addition and subtraction concepts. It is necessary to include addition and subtraction concepts in both books because these ideas are critical to both grade level bands. There may even be situations where both books are needed. Some students may need to begin with the early number concepts. Other students may start with addition and subtraction concepts. The point is to use the formative assessments to find out exactly what the students need . . . and let the games begin!

______________'s Math Progress

Date/s: ______________

Concept: ______________

Things I know about this concept:

____

____

Things I have done with this concept:

____

____

Goals:

____

____

____

How much I know about this concept;

(place X on the scale with date underneath to show level of knowledge; 10 = complete knowledge)

0 ●━━━━━━━━━━━━━━━━━━━━━━━━━━━━━━━━━● 10

CHAPTER

Addition and Subtraction Concepts

Total and Parts
Counting On and Counting Back
Joining Sets
Number Line Proficiency
Take Away Subtraction
Missing Part Subtraction
Comparison Subtraction
Adding and Subtracting Tens
Adding Doubles and Near Doubles
Fact Families—Addition and Subtraction
Partial Sums
Partial Differences
Near Tens for Addition and Subtraction
Equal Differences

CONCEPT:

Total and Parts

What is the Total and Parts Concept?

The total and parts concept encourages students to see the parts of whole numbers. By focusing on total and parts, students learn about addition and subtraction. They begin to understand that the parts make the total and that the total is made of parts. The beauty of teaching total and parts is that we can present early addition and subtract concepts at the same time rather than teaching these operations separately.

CCSS

Operations and Algebraic Thinking
Number and Operations in Base Ten

Formative Assessment

To find out if a student understands total and parts, ask the student to describe the total and parts of the following trains of cubes:

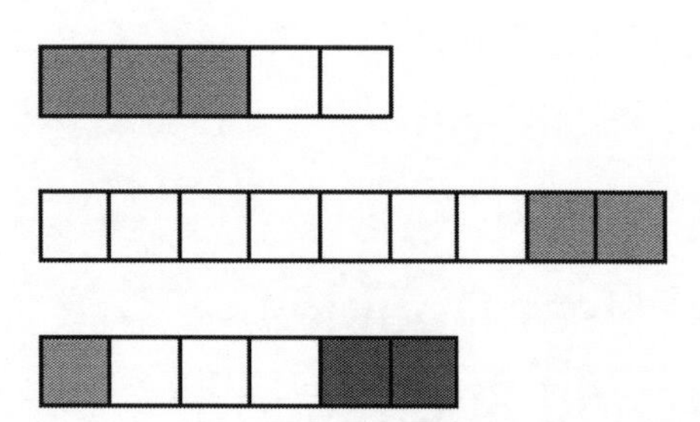

Notice if the student provides the total and parts in the descriptions. If the student is successful, try different parts and larger totals. Identify which totals and parts the students know and which they do not know.

Successful Strategies

Interchange addition and subtraction ideas as students work with total and parts. Encourage students to read aloud and record the parts and total. Using different colors to show the parts often helps students better identify the parts. Grouping the parts is also beneficial. As students gain comfort with total and parts, begin working with more than two parts of each total.

Math Words to Use

total, whole, parts, addition, subtraction

Questions at Different *Levels of Cognitive Demand*

EASY

Recall: *What is the total?*

Comprehension: *How do the parts relate to the total?*

Application: *How could you show different parts of the total?*

Analysis: *How could you compare the parts of more than one total?*

Evaluation: *What helps you understand the parts of a total?*

COMPLEX

Synthesis: *What if the total had many parts?*

Rigorous Problem Solving with the Concept of Total and Parts 1

The teacher gave each student 12 stickers.

David placed 3 stickers on his folder, 5 stickers on his agenda, and some stickers on the class chart.

Sandy placed 8 stickers on her folder, some stickers on her agenda, and 3 stickers on the class chart.

Eugene placed some stickers on his folder, 4 stickers on his agenda, and 2 stickers on the class chart.

Part A

Which equations represent the students' stickers?

- o $3 + 5 = 8$
- o $3 + 5 + 2 + 2 = 12$
- o $3 + 5 + 4 = 12$
- o $8 + 3 = 11$
- o $8 + 1 + 3 = 12$
- o $6 + 4 + 2 = 12$
- o $4 + 2 = 6$
- o $6 + 6 = 12$
- o $7 + 3 + 2 = 12$

Part B

How many stickers were placed on the class chart?

How many stickers were placed on the agendas?

How many stickers were placed on the folders?

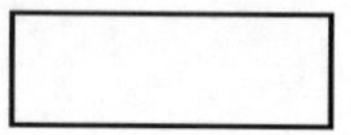

Justify your answers with equations.

Rigorous Problem Solving with the Concept of Total and Parts 2

The teacher gave each student 10 stickers.

Kassidy placed 2 stickers on her folder, 6 stickers on her agenda, and some stickers on the class chart.

Danielle placed 5 stickers on her folder, some stickers on her agenda, and 2 stickers on the class chart.

Samantha placed some stickers on her folder, 3 stickers on her agenda, and 4 stickers on the class chart.

Part A

Which equations represent the students' stickers?

- o $5 + 5 = 10$
- o $5 + 2 = 7$
- o $2 + 6 + 1 + 1 = 10$
- o $2 + 6 + 2 = 10$
- o $3 + 3 + 4 = 10$
- o $2 + 6 = 8$
- o $3 + 3 + 4 = 10$
- o $5 + 3 + 2 = 10$
- o $6 + 6 = 12$

Part B

How many stickers were placed on the class chart? 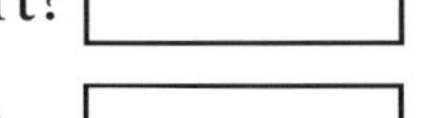

How many stickers were placed on the agendas? 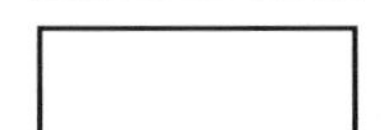

How many stickers were placed on the folders?

Justify your answers with equations.

Big Top Twelve

A game designed to build the concept of total and parts

Materials:

- Die (labeled 1–6 with numerals)
- Crayons or colored pencils
- Big Top Twelve Game Board for each player

Directions:

1. The first player rolls the die. The number showing is one part of twelve.
2. The player decides which Big Top she wants to use to show the part. The player colors the part with a specific color. For example, if the player rolls five, she colors five rectangles (next to one another) in one of the Big Tops with a black crayon.

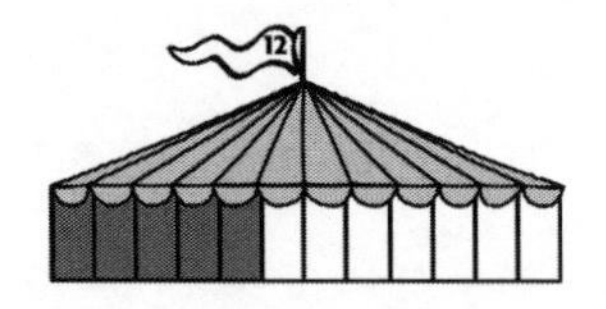

3. The next player rolls and colors the part of twelve on her game board.
4. Players take turns rolling and coloring parts of twelve. Each new part of the Big Top is shown using a different color.
5. The number rolled cannot be split. If the number rolled cannot be played, the player loses her turn.
6. The first person to complete all three Big Tops is the winner.

Content Differentiation:

Moving Back: Try Big Tops that are less than 12.

Moving Ahead: Play Big Top Twelve without the game board. Each time a number is rolled, it becomes part of one of three addition equations. For example, $2 + 3 + 5 + 2 = 12$. Try other Big Top values.

Big Top Twelve Game Board

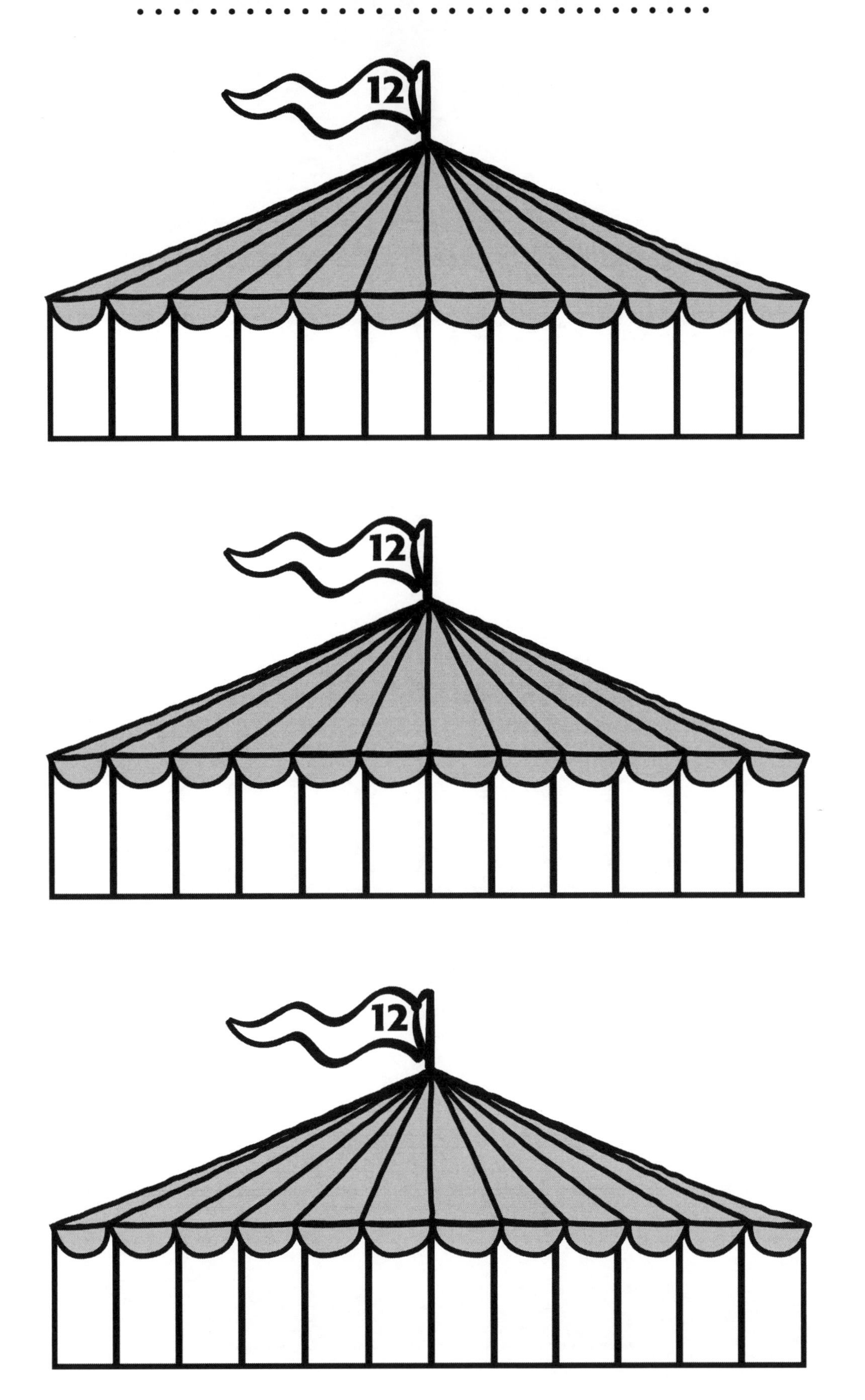

Big Top Twelve Game Board

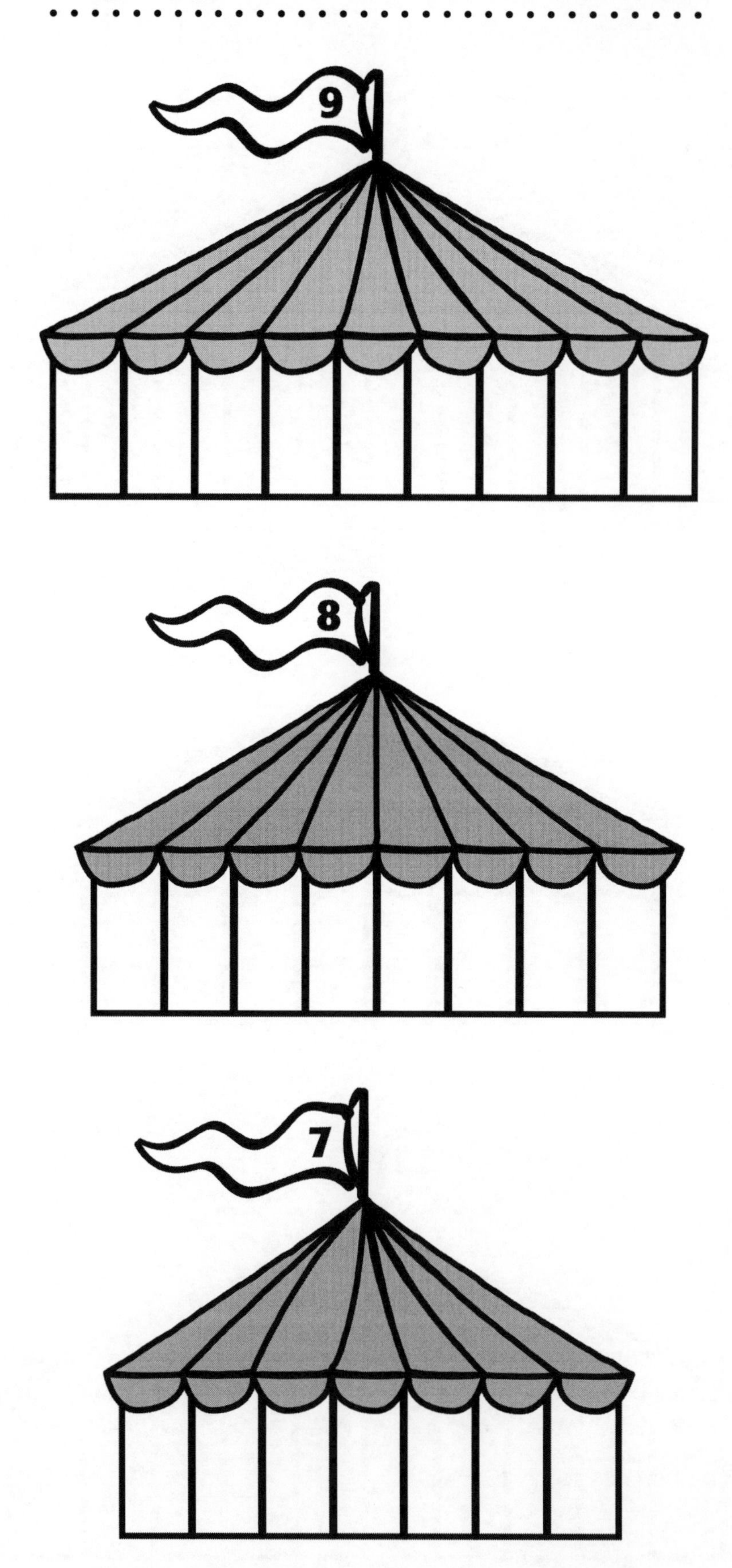

Big Top Twelve Game Board

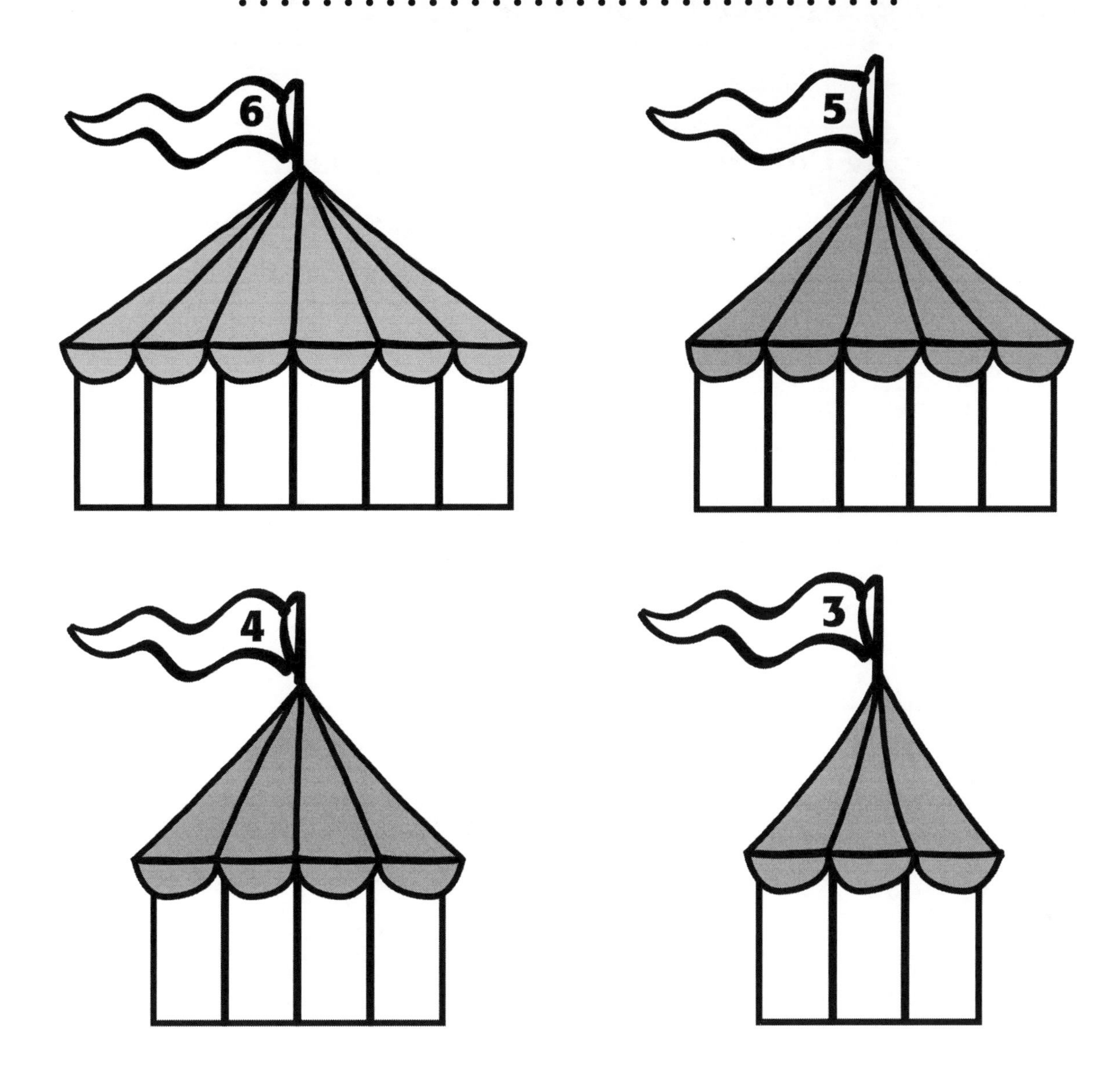

CONCEPT:

Counting On and Counting Back

What are Counting On and Counting Back?

Counting on and counting back are strategies that help students with addition and subtraction. Counting on and counting back are efficient strategies when the addition equation is +1, +2, or +3 and when the subtraction equation is –1, –2, or –3. In a situation such as 48 + 23, counting on by ones is not the most efficient strategy because it is too time consuming. Students need to know how to count on and count back, but they also need to judge when it is a good idea to count on or count back.

CCSS

Counting and Cardinality
Operations and Algebraic Thinking
Number and Operations in Base Ten

Formative Assessment

To find out if a student can count on and count back ask the student to name one more than several given numbers (23, 57, 79). Then try two more, three more, one less, two less, and three less.

23 and one more	57 and one more	79 and one more
23 and two more	57 and two more	79 and two more
23 and three more	57 and three more	79 and three more
One less than 23	One less than 57	One less than 79
Two less than 23	Two less than 57	Two less than 79
Three less than 23	Three less than 57	Three less than 79

Identify which situations the students answer correctly and how quickly the students respond, to determine how comfortable the students are with counting on and counting back by ones.

Successful Strategies

Identifying when it is efficient and when it is not efficient to use counting on and counting back is just as important as knowing how to count on and count back. When students know how and when to count on by ones, we can introduce counting on by numbers other than one. As discussed previously, counting on by ones is not an efficient strategy to use to solve $48 + 23$; although it would be efficient to count on by tens and then ones to solve the equation ($48 + 23 = 48 + 10 + 10 + 1 + 1 + 1$). Similarly, students can count on by other numbers when they are ready.

Math Words to Use

count on, count back, one more, two more, three more, one less, two less, three less

Questions at Different *Levels of Cognitive Demand*

EASY

Recall: *What is one more?*

Comprehension: *What does counting back mean?*

Application: *When is it a good idea to count on by ones?*

Analysis: *How are counting on and counting back similar?*

Evaluation: *When and why is counting back a good strategy?*

Synthesis: *How could you use skip counting by fives to count on?*

COMPLEX

Rigorous Problem Solving with the Concept of Counting On and Counting Back 1

In the game *Nifty Fifty* players can add or subtract the numbers rolled. At the end of the game, the player who scores 50 or is the closest to 50 is the winner.

David had a score of 48. In his last 3 turns, he rolled 4, 5, 3.

Kolby had a score of 52. In his last 3 turns, he rolled 2, 4, 1.

Part A

Which of the following describe how the game ended?

o David won with a score of 51.
o David lost with a score of 53.
o David won with a score of 50.
o Kolby won with a score of 50.
o Kolby lost with a score 49.
o Kolby won with a score of 49.
o Kolby won with a score of 52.
o David and Kolby both scored 49.

Part B

Use words or equations to explain how you know your answer is correct.

Rigorous Problem Solving with the Concept of Counting On and Counting Back 2

In the game *Nifty Fifty* players can add or subtract the numbers rolled. At the end of the game, the player who scores 50 or is the closest to 50 is the winner.

Kirkland had a score of 47. In his last 3 turns, he rolled 3, 1, 3.

Joey had a score of 54. In his last 3 turns, he rolled 2, 4, 2.

Part A

Which of the following describe how the game ended?

- o Kirkland won with a score of 51.
- o Kirkland lost with a score of 48.
- o Kirkland won with a score of 50.
- o Joey won with a score of 50.
- o Joey lost with a score 49.
- o Joey won with a score of 49.
- o Joey won with a score of 52.

Part B

Use words or equations to explain how you know your answer is correct.

Nifty Fifty

A game designed to build the concepts of counting on and counting back

Materials:

- Die (labeled 1, 1, 2, 2, 3, 3)
- Nifty Fifty Score Sheet for each player

Directions:

1. All players record 50 in the "start" column of the first round on their score sheets.
2. The first player rolls the die and records the rolled number on his score sheet. The player chooses whether to count on or count back the number rolled and records the "end" number.
3. The next player rolls, records on her score sheet, and counts on or counts back to find the "end" number.
4. The "end" number becomes the start number for each new round.
5. The player who has 50 after the 10th round is the winner.
6. If no players have 50 after the 10th round, the player who is closest to 50 is the winner.

Content Differentiation:

Moving Back: Change the start/goal number to 20. Write the equation for each round.

Moving Ahead: Play Nifty Fifty without the score sheet. Mentally keep track of "start" and "end" numbers. Try greater start/goal numbers or add more rounds.

Nifty Fifty Score Sheet

Round	Start	Rolled	End
1st			
2nd			
3rd			
4th			
5th			
6th			
7th			
8th			
9th			
10th			

CONCEPT:
Joining Sets

What is the Joining Sets Concept?

Joining sets is an addition concept. Joining sets involves combining quantities and finding the total. Each set that we join is considered an addend. The total of the combined sets is the sum. For some students, the introduction of symbols (+, =), equation formats, and rules come before they have established an understanding of what it means to join sets. These students often struggle with addition because they are trying to remember symbols, formats, and rules that are meaningless to them. It is critical that we help students develop a true understanding of joining sets so that the symbols, formats, and rules make sense to them.

CCSS

Operations and Algebraic Thinking
Number and Operations in Base Ten

Formative Assessment

To find out if a student can join sets use the following number stories. Ask students to explain how they figured out each answer with words, pictures, or models.

Michael has three cookies. Carlos gives him two more cookies. How many cookies does Michael have now?

Gloria has large books and small books on her shelf. There are four large books and five small books. How many books are on Gloria's shelf?

Ten girls line up. Four boys line up. How many students are in the line?

Successful Strategies

To help build students' conceptual knowledge, provide them with models and meaningful situations. It is important that they visualize, manipulate, and talk about joining sets. Teachers do not need to be the sole providers of number stories. Students can and should create story problems as they learn what it means to join sets. Additionally, students need to understand that many sets can be joined. Misconceptions may occur for students if we only provide addition problems that have two addends.

Math Words to Use

join, combine, add, addition, plus, sets, quantities, amounts

Questions at Different *Levels of Cognitive Demand*

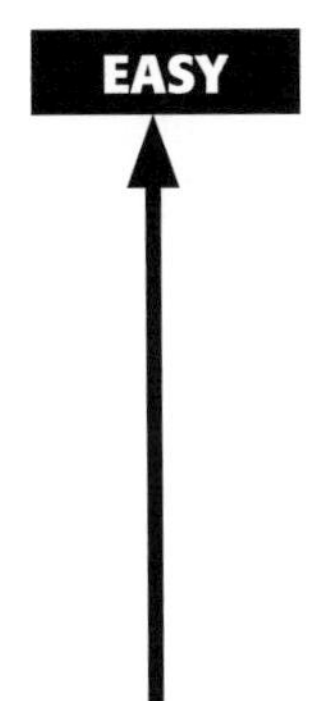

Recall: *What is four and three more?*

Comprehension: *How would you summarize what it means to join sets?*

Application: *How would you show four plus three?*

Analysis: *How are 2 + 3 and 4 + 1 alike?*

Evaluation: *What is the best way to join these sets?*

Synthesis: *How could you rearrange 5 + 2 + 5 + 3?*

Rigorous Problem Solving with the Concept of Joining Sets 1

12 students are playing on the playground at Kayla's school. Some students are on the swings. Some students are on the slide. Some students are jumping rope. Some students are playing soccer.

Part A

Write 4 different equations that could represent the 12 students playing on the playground.

Part B

Use an equation or table to represent the 12 students on the playground if twice as many students are on the swings as are on the slide, and the same number of students are jumping rope as are playing soccer.

Rigorous Problem Solving with the Concept of Joining Sets 2

17 students are playing on the playground at Kayla's school. Some students are on the swings. Some students are on the slide. Some students are jumping rope. Some students are playing soccer.

Part A

Write 4 different equations that could represent the 17 students playing on the playground.

Part B

Use an equation or table to represent the 17 students on the playground if twice as many students are on the swings as are on the slide, and the same number of students are jumping rope as are playing soccer.

Domino Dozen

A game designed to build the concept of joining sets

Materials:

- Set of Double-Six Dominoes

Directions:

1. Place dominoes face down so that dots cannot be seen. To start the game, flip one domino at random.
2. The first player flips a domino and joins the sets of dots on both dominoes.
3. If the total number of dots is a dozen (twelve), the player takes both dominoes. If the total number of dots is not a dozen, the player leaves the dominoes face up.
4. The second player flips one domino and looks for any face-up domino that could be combined with her domino to make a dozen dots. If a combination of twelve can be made, the player takes the dominoes. If a combination of twelve cannot be made the player leaves the dominoes face up.
5. Play continues until all dominoes have been flipped over.
6. The winner is the player with the most dominoes.

Content Differentiation:

Moving Back: Use cubes to represent the number of dots on the dominoes. Or focus on joining the sets of dots on individual dominoes, looking for combinations of six.

Moving Ahead: Write equations for each flip, for example. Or look for combinations of 18 using double-nine dominoes.

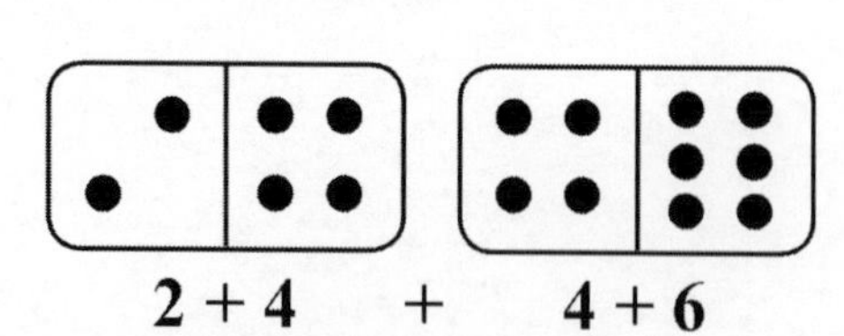

CONCEPT:

Number Line Proficiency

What is Number Line Proficiency?

Number line proficiency involves understanding the order and spacing of numbers assigned to points on a line. Even though every point on a number line has a unique real number (including fractions of fractions), students need to begin with number lines that have the counting by ones points labeled before moving to more complex labels. Number line proficiency includes knowing that on a horizontal number line, movement to the right represents addition and movement to the left represents subtraction. In addition to horizontal number lines, students should also have experiences with vertical number lines. Experiences with number lines that are separated into line segments, such as hundred charts and calendars, and circular number lines (technically not lines), such as analog clocks are also beneficial. Using number lines helps students build understanding of computation and number relationships.

CCSS

Operations and Algebraic Thinking
Number and Operations in Base Ten

Formative Assessment

To find out if a student has number line proficiency, show parts of various number lines and see if students can fill in the missing numbers. Notice if the student quickly identifies the correct missing numbers. Look for patterns in strategies and in errors.

25 26 27 29

0	1		3		5	6		8	9
10	11	12	13				17		19
	21				25			28	

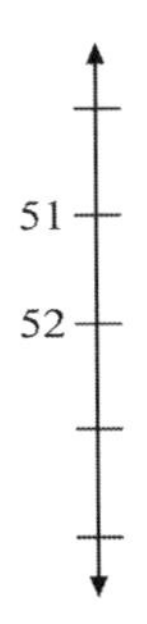

Successful Strategies

For some students, the process of cutting out the rows of a hundred chart and taping the ends together helps them build understanding. They see how the number line is separated into line segments on the hundred chart. This same process can also be used with the calendar.

Math Words to Use

number line, order, count, before, after

Questions at Different *Levels of Cognitive Demand*

EASY

Recall: *What number comes next?*

Comprehension: *What are some examples of numbers that come before?*

Application: *How could you use the number line to show this math situation?*

Analysis: *What numbers go between other numbers on the number line?*

Evaluation: *Which number line works best? Why?*

Synthesis: *How would you design a number line that overlaps with another number line?*

COMPLEX

Rigorous Problem Solving with the Concept of Number Line Proficiency 1

Number line mix up!

Part A

Find and fix the errors to complete the number line diagram.

Hint: 70 is correct. Count by ones.

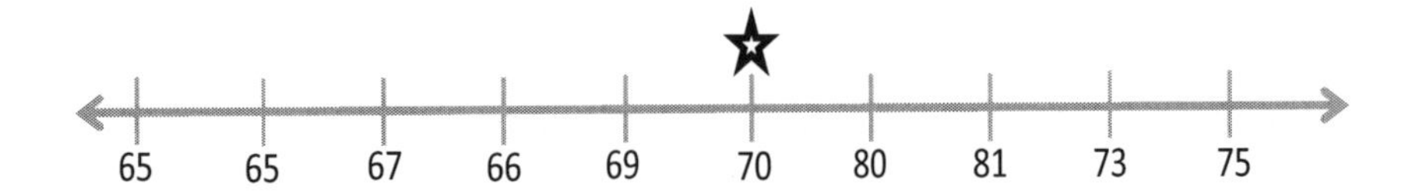

Part B

Correctly place the following numbers to show skip counting on the number line diagram.

100 140 70 150 80 130 90 110 60 120

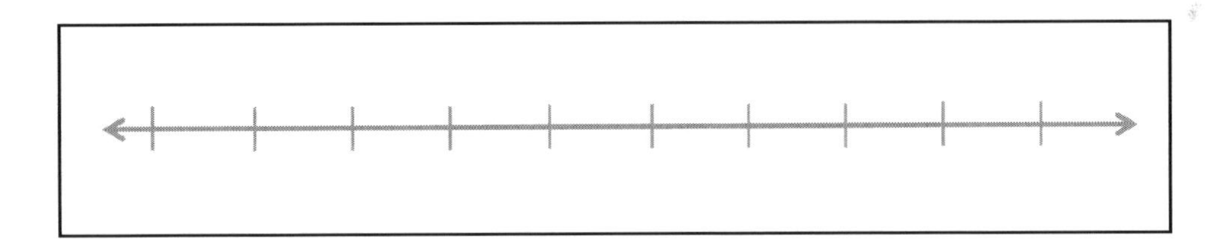

Describe the skip counting.

Rigorous Problem Solving with the Concept of Number Line Proficiency 2

Number line mix up!

Part A

Find and fix the errors to complete the number line diagram.

Hint: 80 is correct. Counts by ones.

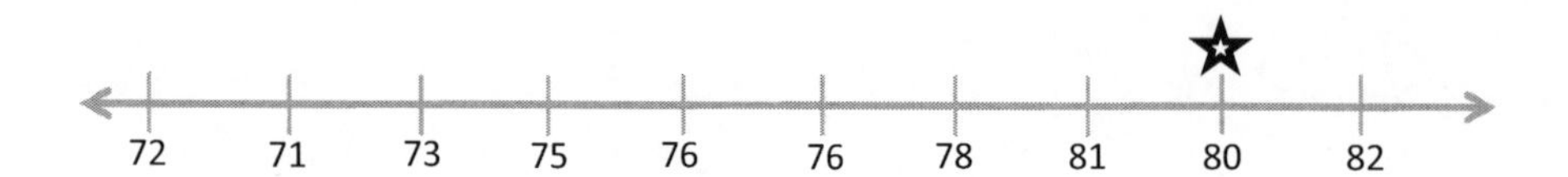

Part B

Correctly place the following numbers to show skip counting on the number line diagram.

55 75 40 65 45 70 80 50 35 60

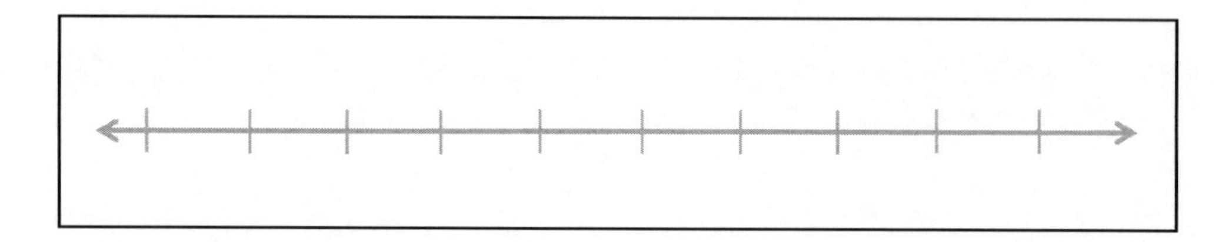

Describe the skip counting.

Number Line Fill Up

A game designed to build the concept of number line proficiency

Materials:

- Number Line Fill Up template for each player
- Decahedron die (ten-sided, labeled 0–9) or 0–9 spinner *(see page 142)*

Directions:

1. Each player rolls the decahedron die (or spins the 0–9 spinner) one time to find his start number.
2. Players decide where to write their start numbers on the Number Line Fill Up template.
3. All other rolls (or spins) are numbers shared by all players.
4. Each player records the new numbers one at a time. In this game, players are rolling (or spinning) one digit at a time to make 2-digit numbers in a counting by ones sequence. If a number cannot be used, the player waits for the next roll (or spin). Once a number is written, it cannot be moved.
5. The winner is the first player to complete an accurate number line chunk including four 2-digit numbers in order.

Content Differentiation:

Moving Back: Show values with base ten blocks or cubes as students build the number line, or try the game using single-digit numbers.

Moving Ahead: Add a third section to each number box and try the game with 3-digit numbers, or try a skip counting sequence on the number line.

Number Line Fill Up

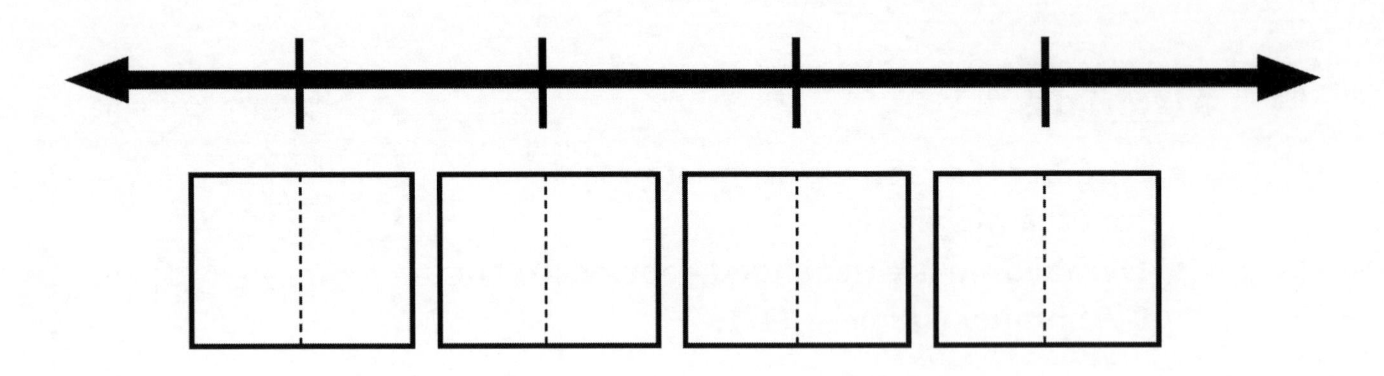

Number Line Fill Up

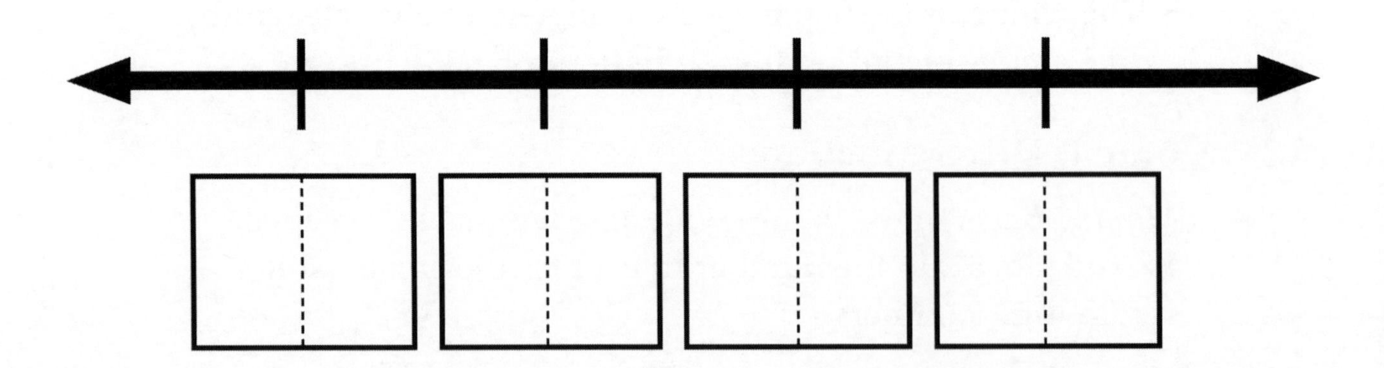

CONCEPT:
Take Away Subtraction

What is the Take Away Subtraction Concept?

Take away subtraction is one of three types of subtraction. Take away subtraction involves set removal. In number stories, a certain amount is removed, eaten, given, lost, hidden, or otherwise taken away. In a subtraction equation, the subtrahend is being subtracted from the minuend to find the difference. In the example $7 - 2 = 5$, 7 is the minuend, 2 is the subtrahend, and 5 is the difference. Students need many meaningful experiences to build conceptual understanding of take away subtraction.

CCSS

Operations and Algebraic Thinking
Number and Operations in Base Ten

Formative Assessment

To find out if a student understands take away subtraction, ask the student to tell a story that matches the following equation: $10 - 4 = 6$. Notice if the student uses words that represent take away subtraction. If the student is not successful, try a smaller equation such as $4 - 1 = 3$ and invite the student to use manipulatives to demonstrate as he tells the number story. If the student is successful with $10 - 4 = 6$, give a larger equation such as $25 - 6 = 19$ and invite the student to write the equation and tell a related number story.

Successful Strategies

The take away subtraction concept is best understood in context. Students need to engage in many types of take away situations to build conceptual understanding. Our goals are not for students to memorize vocabulary such as minuend and subtrahend. Instead we need to model these words as we teach the students the true meaning of take away subtraction.

Math Words to Use
subtract, subtraction, take away, minus sign, equation, minuend, subtrahend, difference

Questions at Different *Levels of Cognitive Demand*

Recall: *What is eight take away three?*

Comprehension: *How would you explain take away subtraction?*

Application: *How would you use take away subtraction in this situation?*

Analysis: *Which number story represents take away subtraction?*

Evaluation: *Which type of subtraction best represents this math situation?*

Synthesis: *If you started with 6 more, how would that affect the difference?*

COMPLEX

Rigorous Problem Solving with the Concept of Take Away Subtraction 1

There were 40 flies. On Monday the hungry frog ate 5 flies. On Tuesday, the hungry frog ate 3 flies. On Wednesday, the hungry frog ate 4 flies. On Thursday, the hungry frog ate 6 flies. On Friday, the hungry frog ate 2 flies. On Saturday, the hungry frog ate 5 flies. On Sunday, the hungry frog ate 1 fly.

Part A

Write an equation to represent how many flies were left after the hungry frog ate on Tuesday.

Write an equation to represent how many flies were left after the hungry frog ate on Friday.

Part B

If the hungry frog eats half as many flies the next week, how many flies will be left?

Justify your answer with one or more equations.

Rigorous Problem Solving with the Concept of Take Away Subtraction 2

There were 50 flies. On Monday the hungry frog ate 2 flies. On Tuesday, the hungry frog ate 4 flies. On Wednesday, the hungry frog ate 3 flies. On Thursday, the hungry frog ate 5 flies. On Friday, the hungry frog ate 2 flies. On Saturday, the hungry frog ate 4 flies. On Sunday, the hungry frog ate 2 flies.

Part A

Write an equation to represent how many flies were left after the hungry frog ate on Wednesday.

Write an equation to represent how many flies were left after the hungry frog ate on Saturday.

Part B

If the hungry frog eats the same amount of flies the next week, how many flies will be left?

Justify your answer with one or more equations.

Hungry Frog

A game designed to build the concept of take away subtraction

Materials:

- Hungry Frog Game Board
- 40 small counters
- Dice (two six-sided cubes labeled 1–6 with numerals)

Directions:

1. Players place one counter on each fly on the game board.
2. The first player rolls both dice and decides which number she wants to use to represent how many flies Hungry Frog will eat. The player tells a number story about Hungry Frog eating flies, models the take away subtraction by taking away counters, and states the subtraction equation. For example, there are 40 flies. Hungry Frog eats three. Now there are 37 flies left, $40 - 3 = 37$.
3. The next player rolls the dice and chooses one of the numbers to be the subtrahend. The player tells the new subtraction story and states the equation. For example, $37 - 5 = 32$.
4. Players take turns rolling the dice, choosing a subtrahend, telling and modeling the associated number story, and stating the equation.
5. The winner is the player who has Hungry Frog eat the last fly. An exact number must be rolled to win.

Content Differentiation:

Moving Back: Use half of the game board and start with 20 flies.

Moving Ahead: Use mental math to play this game without counters or the game board. Tell different kinds of number stories.

Hungry Frog Game Board

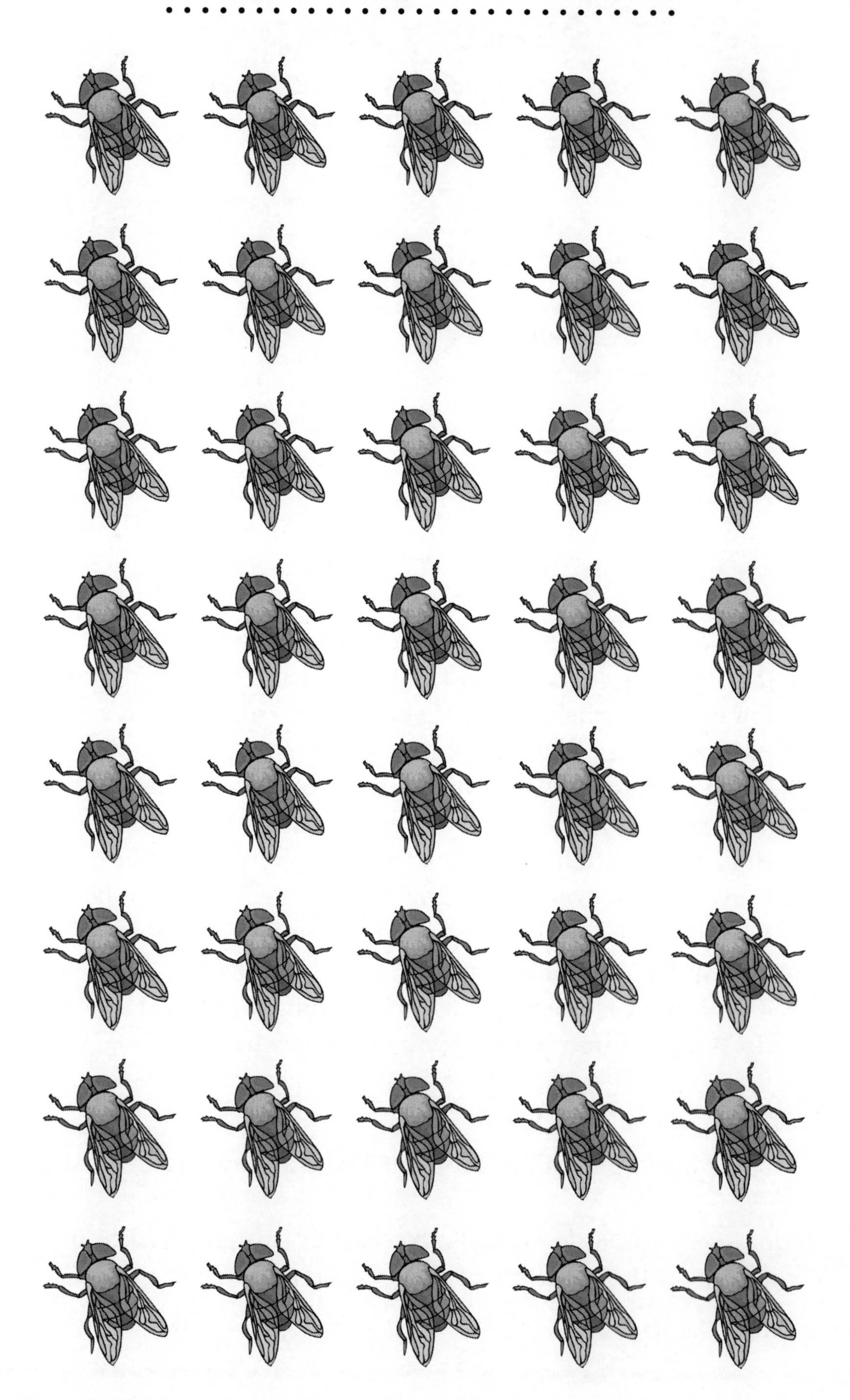

CONCEPT:
Missing Part Subtraction

What is Missing Part Subtraction?

Missing part subtraction is one of three types of subtraction. Missing part subtraction involves an unknown part of the whole. Either the subtrahend or the difference is unknown. For example, $10 - \square = 2$ or $10 - 8 = \square$. Missing part subtraction is often used to solve start unknown or change unknown addition problems. For example, $\square + 2 = 10$ or $8 + \square = 10$. When working with missing part subtraction, it is important that students use "subtract" rather than "take away" to more accurately describe the mathematics situation.

CCSS

Operations and Algebraic Thinking
Number and Operations in Base Ten

Formative Assessment

To find out if a student understands missing part subtraction, ask the student to solve the following number stories;

Melanie has 5 bracelets in her jewelry box. She puts some of the bracelets on her wrist. There are 2 bracelets left in her jewelry box. How many bracelets are on her wrist?

An apple tree has 11 beautiful apples. The wind blows 4 apples off the tree. How many apples are still on the tree?

Lenny bakes 24 cupcakes. He packs a lot of the cupcakes in a box and leaves 6 on the table. How many cupcakes are in the box?

Successful Strategies

To help students understand the concept of missing part subtraction use models to show the parts of the total. Color coding parts and arranging the parts in groups are helpful ways to highlight the parts within the whole set.

Math Words to Use

subtract, subtraction, difference, minuend, subtrahend, minus sign, missing part

Questions at Different *Levels of Cognitive Demand*

EASY

Recall: *What is $6 - 2$?*

Comprehension: *How do you know that $10 - 3 = 7$?*

Application: *Is there another number story that could go with this subtraction equation?*

Analysis: *Which number stories represent missing part subtraction?*

Evaluation: *What is the best way to model missing part subtraction?*

Synthesis: *How could you reorganize these facts to show a new pattern?*

COMPLEX

Rigorous Problem Solving with the Concept of Missing Part Subtraction 1

Teri has 18 pennies. She leaves some on the table and hides some in her hand.

Complete the chart to represent how many pennies are on the table, how many are hiding in her hand, and the related subtraction equations. *Note: Teri does not hide or leave zero pennies on the table.*

Part A

Total Pennies	On the Table	Hiding in Teri's hand	Subtraction Equation
18	6		$18 - 6 = 12$
18		10	
	13	5	
18	7		
18		15	
	4	14	
18	2		

Part B

Write subtraction equations to represent two other ways Teri could hide part of the 18 pennies.

Rigorous Problem Solving with the Concept of Missing Part Subtraction 2

Donald has 20 pennies. He leaves some on the table and hides some in his hand.

Complete the chart to represent how many pennies are on the table, how many are hiding in his hand, and the related subtraction equations. *Note: Donald does not hide or leave zero pennies on the table.*

Part A

Total Pennies	On the Table	Hiding in Donald's hand	Subtraction Equation
20	15		$20 - 15 = 5$
20		10	
	13	7	
20	17		
20		12	
	6	14	
20	19		

Part B

Write subtraction equations to represent three other ways Donald could hide part of the 20 pennies.

Broken Towers

A game designed to build the concept of missing part subtraction

Materials:

- Linking cubes (20 for each player)

Directions:

1. The first player states a number (1–20).
2. All players build a tower using the stated number of cubes.
3. When the towers are completed, each player holds her tower under the table. The players break their towers into two parts.
4. One at a time, players show one part of their tower by placing it on the table. The other players count the cubes in the part of the tower that is showing and try to name how many cubes are in the missing part under the table.
5. The first player to name the correct number of cubes in the missing part of the tower earns one point. The player showing the tower can also earn one point by successfully stating the missing part subtraction equation.
6. After all players have shown their towers, the round is complete. A new number is given for the next round.
7. The winner is the player with the most points after five rounds.

Content Differentiation:

Moving Back: Build towers of ten or less.

Moving Ahead: Break towers into three parts.

CONCEPT:
Comparison Subtraction

What is Comparison Subtraction?

Comparison subtraction is one of three types of subtraction. In comparison subtraction, the focus is on the difference between quantities. Comparing means finding out how many more or how many less one number is compared to another number. As with missing part subtraction, comparison subtraction is not "take away." Finding the difference does not involve removal; therefore we should call it subtraction rather than "take away." Differences are reported in absolute values. This means that we do not need to teach students rules such as *the big number goes first when we subtract*. Instead, we can focus on teaching the concept of comparison subtraction because the difference between two numbers can be positive or negative.

CCSS

Operations and Algebraic Thinking
Number and Operations in Base Ten

Formative Assessment

To find out if a student understands comparison subtraction, give the student some numbers to compare. Ask the student to explain how he solved each problem.

What is the difference between 5 and 4?
What is the difference between 10 and 7?
What is the difference between 15 and 9?
What is the difference between 20 and 12?

If the student is successful, try numbers with greater differences. If the student is not successful, encourage the student to use objects to model the differences.

Successful Strategies

Comparison subtraction is difficult for many students. To help highlight differences between quantities, students can use manipulatives. Towers of cubes are especially helpful. When we ask the student to name the difference, we can discuss how many more are needed to make the towers even or equal.

Math Words to Use

subtract, subtraction, compare, comparison, difference, subtrahend, minuend, minus sign

Questions at Different *Levels of Cognitive Demand*

EASY

Recall: *What is the difference between 15 and 5?*

Comprehension: *How would you explain comparison subtraction?*

Application: *How would you show 12 – 4 with pictures?*

Analysis: *Which number stories represent comparison subtraction?*

Evaluation: *What is the best way to find the difference? Why is it best?*

Synthesis: *How would you design a new template for comparison subtraction?*

COMPLEX

Rigorous Problem Solving with the Concept of Comparison Subtraction 1

The Cousin Family went fishing. The chart shows how many fish each member of the family caught.

Cousin Family Members	Number of Fish
Joey	12
Kolby	9
James	4
Kirkland	15
Kayla	8

Part A

The difference between the number of fish Joey caught and the number of fish James caught is ☐.

The related comparison subtraction equation is ☐.

The difference between the number of fish Kirkland caught and the number of fish Kolby caught is ☐.

The related comparison subtraction equation is ☐.

Part B

Which family members' fish could be compared to find a difference of 3?

Is there more than one answer? Circle Yes or No. Write the equation(s).

Rigorous Problem Solving with the Concept of Comparison Subtraction 2

The Paw Family went fishing. The chart shows how many fish each member of the family caught.

Paw Family Members	Number of Fish
Lilly	5
Beau	11
Luna	7
Mac	16
Xina	9

Part A

The difference between the number of fish Beau caught and the number of fish Xina caught is ☐.

The related comparison subtraction equation is ☐.

The difference between the number of fish Mac caught and the number of fish Luna caught is ☐.

The related comparison subtraction equation is ☐.

Part B

Which family members' fish could be compared to find a difference of 11?

Is there more than one answer? Circle Yes or No. Write the equation(s).

Compare Bear

A game designed to build the concept of comparison subtraction

Materials:

- Number Mat for each player
- Two different colored bear counters for each player

Directions:

1. The first player drops the two bears onto the Number Mat. The player announces the two numbers that the bears land on (or are closest to) and states the difference. For example, if the bears land on 14 and 8, the player says, "The difference between 14 and 8 is 6."
2. The player also tells a comparison subtraction story. For example, "Blue Bear has 14 cookies. Red Bear has 8 cookies. Blue Bear has 6 more cookies than Red Bear."
3. The next player drops two bears on his number mat. The player announces the two numbers, states the difference, and tells a comparison subtraction story.
4. The player with the greater difference scores one point.
5. Players take turns dropping bears on their number mats, stating the difference, and telling the associated number stories.
6. The winner is the first player to score ten points.

Content Differentiation:

Moving Back: Use connecting cubes to show the numbers to help students see the differences.

Moving Ahead: Try the same game using a hundred chart as the number mat.

Number Mat

1	2	3	4	5
6	7	8	9	10
11	12	13	14	15
16	17	18	19	20

CONCEPT:
Adding and Subtracting Tens

What is the Adding and Subtracting Tens Concept?

Adding and subtracting tens builds fluency with number relationships and computation. When students understand the adding and subtracting tens concept, they are more apt to use mental math to solve problems and verify answers. Initially, students add and subtract tens to and from multiples of ten. Building on this idea, they learn to add and subtract tens to and from any number.

CCSS

Operations and Algebraic Thinking
Number and Operations in Base Ten

Formative Assessment

To find out if a student can add and subtract tens, ask the student to solve the following equations:

20 + 10 =
40 – 10 =
50 + 30 =
70 – 20 =
65 + 10 =
82 – 10 =
38 + 40 =
97 – 30 =

Ask the student to explain how he solved each problem. Analyze the student's proficiency through his accuracy, speed, and explanations.

Successful Strategies

One of the foundational concepts associated with adding and subtracting tens is rational counting by tens. Encourage students to skip count by tens beginning with zero and with numbers other than zero. To further help students understand the adding and subtracting tens concept, use base ten manipulatives or drawings to model the computation.

Math Words to Use

tens, add, addition, subtract, subtraction, minus sign, plus sign

Questions at Different *Levels of Cognitive Demand*

Recall: *What is 53 + 10?*

Comprehension: *How do you add ten to a number?*

Application: *How would you use 92 – 10 to solve 92 – 20?*

Analysis: *Which facts go together? Why?*

Evaluation: *How does skip counting help us add and subtract ten?*

Synthesis: *How could you use the adding and subtracting tens concept to add and subtract nines?*

Rigorous Problem Solving with the Concept of Adding and Subtracting Tens 1

Hundred Chart Piece

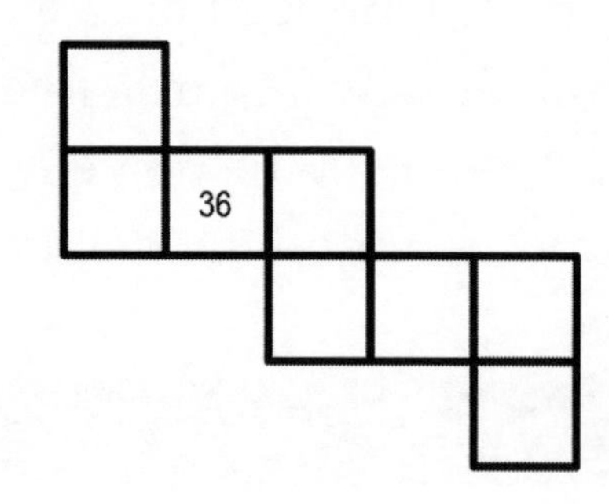

Part A

Which set of numbers fills in all of the missing numbers on the piece of the hundred chart?

o 34, 35, 37, 38, 39, 40, 41
o 24, 25, 37, 38, 39, 40, 50
o 26, 36, 37, 47, 48, 49, 50
o 26, 35, 37, 47, 48, 49, 59
o 25, 35, 37, 47, 48, 49, 59

Part B

How do you know that 57 is <u>not</u> one of the missing numbers?

Justify your answer with at least two equations.

Rigorous Problem Solving with the Concept of Adding and Subtracting Tens 2

Hundred Chart Piece

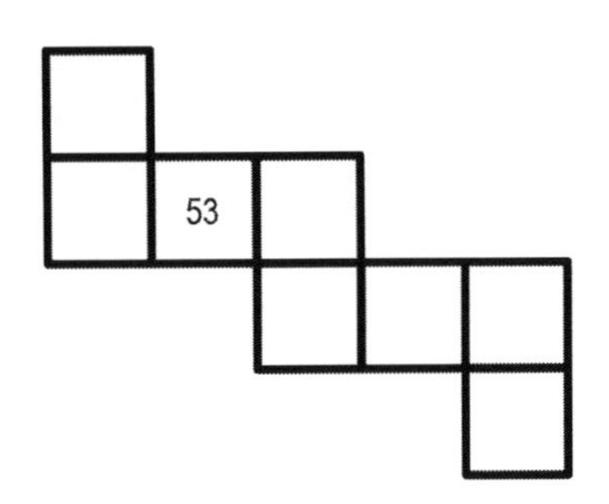

Part A

Which set of numbers fills in all of the missing numbers on the piece of the hundred chart?

- o 51, 52, 54, 55, 56, 57, 58
- o 42, 43, 54, 55, 65, 66, 76
- o 51, 52, 54, 64, 65, 66, 76
- o 42, 52, 54, 64, 65, 66, 76
- o 43, 44, 54, 55, 65, 66, 67

Part B

How do you know that 73 is *not* one of the missing numbers?

Justify your answer with at least two equations.

Hundred Chart Tic-Tac-Toe

A game designed to build the concept of adding and subtracting tens

Materials:

- Markers or crayons (different color for each player)
- Paper

Directions:

1. Each player decides which color marker/crayon to use.
2. One player draws a tic-tac-toe board on the paper.
3. The first player writes a 2-digit number in one of the spaces.
4. The next player uses adding or subtracting ones or tens to write a related number in one of the spaces. The placement of numbers is relative to position on a hundred chart.
5. Players take turns filling in related numbers.
6. The winner is the first player to have three numbers in a horizontal, vertical, or diagonal row. For example:

54	55	56
64	65	66
74		76

Content Differentiation:

Moving Back: Use a hundred chart as a reference. Or draw a large tic-tac-toe board and place base ten blocks on the board to represent the numbers.

Moving Ahead: Make a tic-tac-toe board with more lines and try four in a row.

CONCEPT:
Adding Doubles and Near Doubles

What is the Adding Doubles and Near Doubles Concept?

There is something uniquely fascinating about doubles. Often students are keenly interested in doubles, which propels them to master these facts earlier than other facts. The smaller doubles (1+1, 2+2, 3+3, 4+4, 5+5) are often learned quickly. Larger doubles (6+6, 7+7, 8+8, 9+9) can take more time. To help students learn doubles, provide many experiences with doubling situations. Students also need to conceptually understand addition and the idea of "twice as many." When students own their doubles, meaning they do not need to count all or count on, they are ready to learn how to use this knowledge to solve near double addition problems. When teaching the near doubles concept, encourage students to identify the two doubles facts that the near double falls between and allow them to use plus one or minus one to solve the equation. For example, $7 + 8 = 15$ falls between $7 + 7 = 14$ and $8 + 8 = 16$. Allow the students to decide which double they want to use to solve the near double problem. In this way, students learn more about number relationships. They know they need to add one or subtract one because they understand the math involved.

CCSS

Operations and Algebraic Thinking
Number and Operations in Base Ten

Formative Assessment

Analyze the level of ownership by the accuracy and speed with which the student solves the doubles problems. If the student owns her doubles, provide near doubles problems and ask her to explain how she solved them. Identify which doubles and near doubles the student does and does not own.

Successful Strategies

Teaching students to identify doubles and near doubles is the first step in ownership of these facts. Ask students to explain which two doubles facts "sandwich" the near doubles fact. Use manipulatives (towers of cubes or stacks of counters) to model one more and one less. Write out the facts so that students can see the patterns. As students gain ownership of doubles and near doubles, we need to encourage them to use this knowledge with related subtraction facts.

Math Words to Use

doubles, twice as many, add, addition, facts, sum, near doubles, one more, one less

Questions at Different *Levels of Cognitive Demand*

EASY

Recall: *What is 4 + 4?*

Comprehension: *How would you explain double?*

Application: *How could you use 6 + 6 to solve 6 + 7?*

Analysis: *Why is 5 + 5 similar to 10 + 10?*

Evaluation: *How would you help someone else solve 7 + 8?*

Synthesis: *What if you were to invent a new way to organize the doubles facts?*

COMPLEX

Rigorous Problem Solving with the Concept of Adding Doubles and Near Doubles 1

Samantha and Kayla are playing *X Marks the Spot*. In this game, players roll a number, double it, and mark the sum on the game board. Samantha rolled 5, 3, and 4. Kayla rolled 6, 2, and 5.

Part A

Which set of numbers represents Kayla's sums?

o 12, 4, 5
o 10, 6, 8
o 12, 2, 10
o 11, 4, 10
o 12, 4, 10
o 10, 4, 8
o 12, 4, 11

Part B

James says 7 + 8 = 17 because 8 + 8 = 16 and 16 + 1 = 17
Do you agree with James?

Why or why not?

Rigorous Problem Solving with the Concept of Adding Doubles and Near Doubles 2

Kass and Dani are playing *X Marks the Spot*. In this game, players roll a number, double it, and mark the sum on the game board. Kass rolled 4, 2, and 5. Dani rolled 5, 3, and 4.

Part A

Which set of numbers represents Dani's sums?

o 8, 6, 8
o 10, 6, 4
o 10, 5, 8
o 10, 3, 8
o 8, 4, 10
o 10, 3, 4
o 10, 6, 8

Part B

Kirk says 6 + 7 = 13 because 6 + 6 = 12 and 12 + 1 = 13.
Do you agree with Kirk?

Why or why not?

X Marks the Spot

A game designed to build the concept of adding doubles and near doubles

Materials:

- 0–9 Number Cards or 0–9 Number Tiles in a paper cup
- X Marks the Spot Game Boards
- Two highlighters or light markers (different colors)

Directions:

1. Players decide which color highlighter each person will use.
2. Without looking into the cup, the first player chooses a number. The player doubles the number and looks for the sum on the game board.
3. The player marks the sum with a dot using the highlighter to "hold" the spot and places the number back into the cup.
4. The next player chooses a number, doubles the number and looks for the sum on the game board. The player can mark the sum with a dot to hold it OR the player can "claim" any sum that is currently being held with a dot.
5. If the player claims a held sum, the highlighter is used to draw X on the sum.
6. Players take turns choosing numbers, doubling, and holding or claiming sums. *Note: All sums must be held before they are claimed. Held sums can be claimed by any player.*
7. The winner is the first player to have four claimed sums (Xs) in a row (horizontal, vertical, or diagonal).

Content Differentiation:

Moving Back: Play the game without holding sums prior to claiming sums.

Moving Ahead: Play the same game using Near Doubles. The near double is one more or one less than the closest double.

0–9 Number Cards

0	1	2	3	4
5	6	7	8	9

X Marks the Spot Game Board for Doubles

X Marks the Spot Game Board for Near Doubles

11	9	13	7	15
5	17	3	19	1
1	19	3	17	5
15	7	13	9	11
11	9	13	7	15
5	17	3	19	1

CONCEPT:

Fact Families–Addition and Subtraction

What are Addition and Subtraction Fact Families?

Fact families for addition and subtraction are facts that are directly related. In a fact family, there are two addition problems and two subtraction problems. The addition facts are manifestations of the commutative property. The subtraction facts are inverse operations of the addition facts.

CCSS

Operations and Algebraic Thinking

Formative Assessment

To find out if a student understands addition and subtraction fact families, provide the following incomplete fact family and ask the student which fact is missing from the family.

8 + 4 = 12
12 – 4 = 8
12 – 8 = 4

If the student is not successful, try a smaller fact family. If the student is successful, ask her to generate all of the facts in the 7, 8, 15 fact family.

Successful Strategies

Some students struggle with fact families. One of the ways to help these students is to go back to the parts and total concept. Focusing on the parts of a given total encourages students to think about the related addition and subtraction facts. Additionally, help student see why doubles produce identical facts.

Math Words to Use

addition, subtraction, equation, facts, fact families

Questions at Different *Levels of Cognitive Demand*

Recall: *What is 5 + 2? What is 2 + 5?*

Comprehension: *How would you explain fact families?*

Application: *How would you show the fact family of 6, 7, 13?*

Analysis: *What relationship do you see between 8 + 9 and 17 – 9?*

Evaluation: *What advice would you give to someone who does not understand fact families?*

Synthesis: *Is it possible to have a fact family for this equation 2 + 3 + 4 = 9?*

COMPLEX

Rigorous Problem Solving with the Concept of Fact Families 1

In the Funny Bunny game, players try to name the number hidden on one of the bunny's ears and the four facts in the fact family.

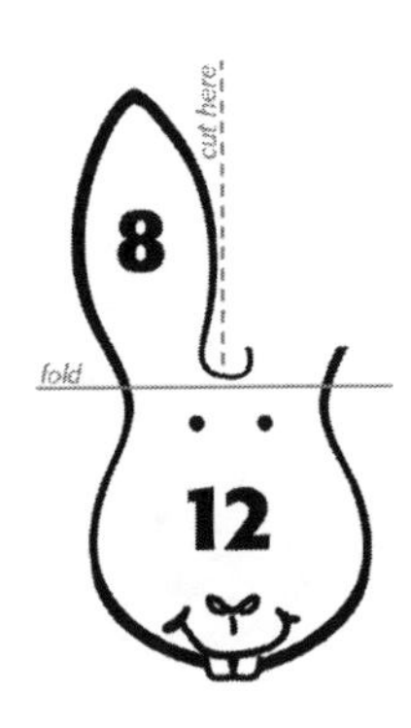

Part A

Which fact belongs to the fact family represented above?

- o 8 + 12 = 21
- o 8 – 12 = 4
- o 4 + 8 = 12
- o 21 – 8 = 12
- o 4 + 12 = 8

Part B

List the four facts in a different fact family that has the sum of 12.

Rigorous Problem Solving with the Concept of Fact Families 2

In the Funny Bunny game, players try to name the number hidden on one of the bunny's ears and the four facts in the fact family.

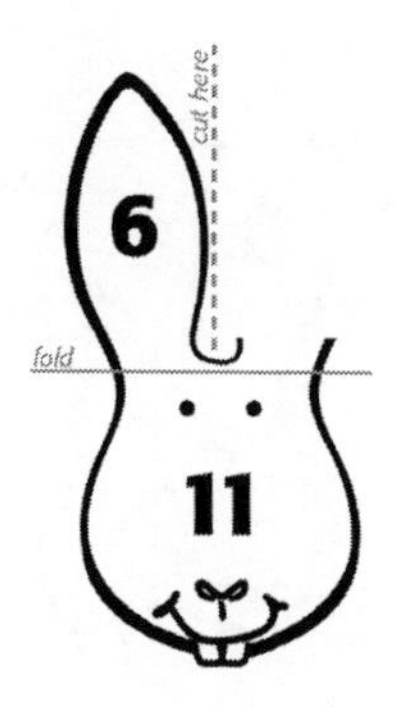

Part A

Which fact belongs to the fact family represented above?

o $6 + 11 = 18$
o $11 - 5 = 6$
o $11 - 6 = 7$
o $18 - 6 = 12$
o $6 + 5 = 12$

Part B

List the four facts in a different fact family that has the sum of 11.

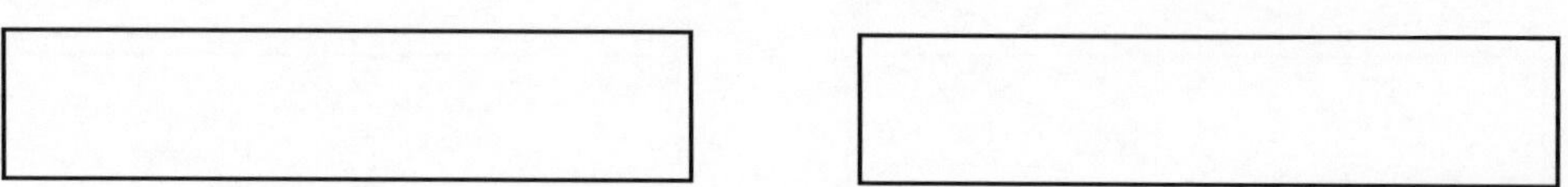

Funny Bunny

A game designed to build the concept of addition and subtraction fact families

Materials:

- Funny Bunny Resource Pages
- Straws
- Tape

Directions:

1. Cut out bunnies and tape each bunny on a straw.
2. Place all bunny puppets face down on the table.
3. The first player chooses a bunny and folds back one of the bunny's ears.
4. The other player names the hidden number on the bunny's ear. The player receives one point for each fact in the fact family that is named correctly.
5. Players take turns choosing bunnies, folding ears, and naming the facts in the fact family.
6. The first player to reach 20 or more points is the winner.

Content Differentiation:

Moving Back: Look at the bunny with both ears up and name the facts in the fact family.

Moving Ahead: Use the blank bunnies to make different fact families.

Funny Bunny Resource Page

Funny Bunny Resource Page

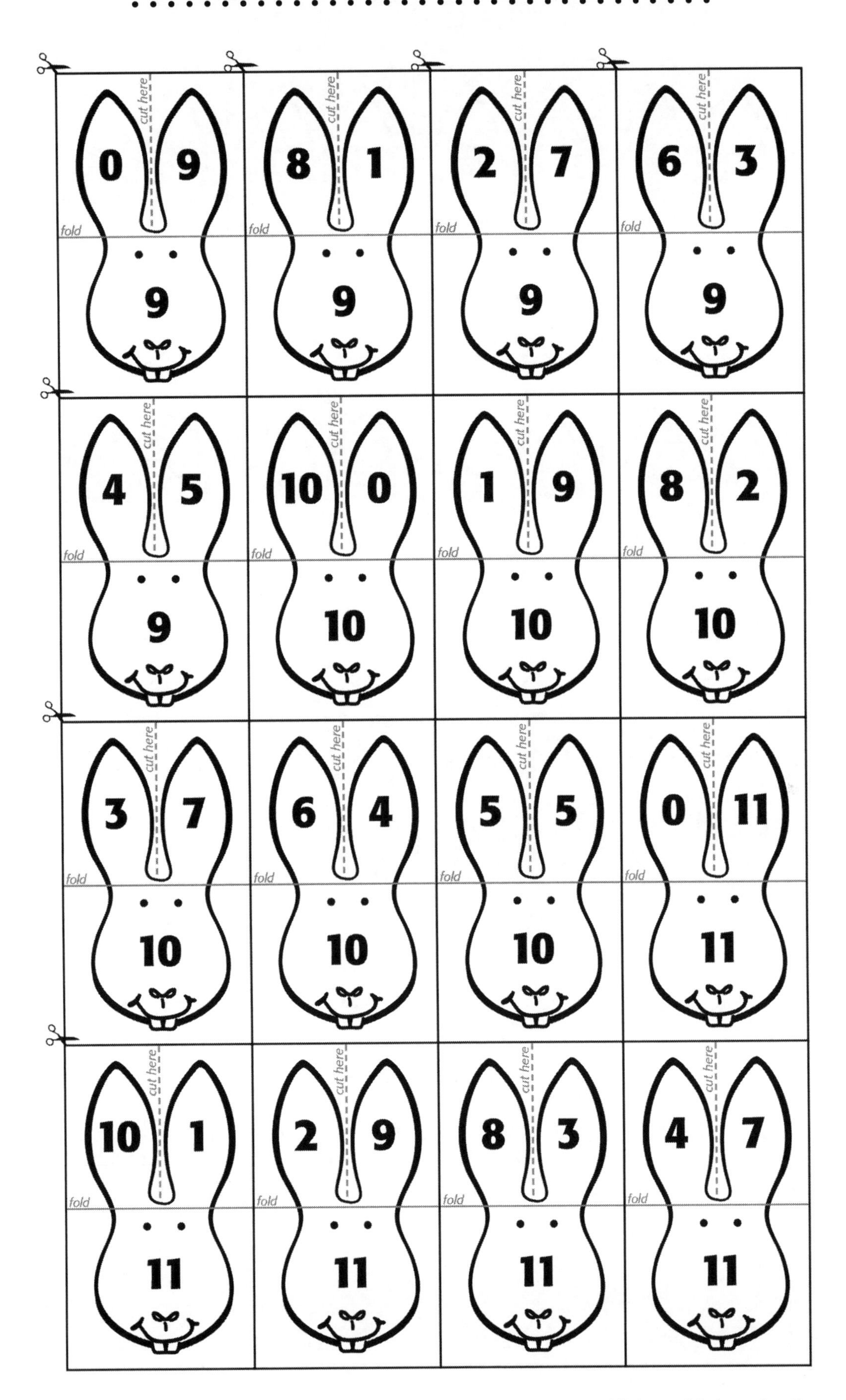

Funny Bunny Resource Page

CONCEPT:
Partial Sums

What is the Partial Sums Concept?

The partial sums concept is an addition strategy that involves combining addends into workable clusters. The traditional algorithm for addition involves following procedures, such as *aligning digits in columns* and *always starting with the ones.* The rules of the traditional algorithm are meaningless for many students. They cannot remember what to do because they do not understand the math. Alternative algorithms such as partial sums build conceptual understanding. With partial sums, it does not matter if columns are aligned or if the student starts with the ones or not. There is no need to "carry." For example, 37 + 48 = 70 + 15. In this situation, the tens are clustered (70 is a partial sum) and the ones are clustered (15 is a partial sum). The partial sums are easily combined into 85. The idea with partial sums is to cluster the addends in a way that invites mental math.

CCSS

Operations and Algebraic Thinking
Number and Operations in Base Ten

Formative Assessment

To find out if a student understands the partial sums concept, ask the student to solve the following equations:

23 + 42 =   26 + 23 =   31 + 39 =
68 + 12 =   57 + 38 =   76 + 17 =

Ask the student to explain how she solves each problem. Invite the student to try the problems without aligning the digits in columns and without starting with the ones.

Successful Strategies

Some of the reasons students struggle with the traditional algorithm include the phrases that they have heard. For example, many people say things like, "Carry the one," when they are actually composing a ten. Because the math language is not accurate, the students develop misconceptions. The partial sums concept is embedded in place value and conceptual understanding, which help remedy algorithmic misconceptions. Even if students successfully use the traditional algorithm, learning the partial sums concept invites students to strengthen conceptual understanding, increase mental math capacity, and improve flexibility and fluency.

Math Words to Use

partial, part, sum, addition, add, strategy, equation, equal

Questions at Different *Levels of Cognitive Demand*

EASY

Recall: *What is 25 + 15?*

Comprehension: *How did you solve the problem?*

Application: *What picture could you draw to show the partial sums?*

Analysis: *How would you compare the partial sums?*

Evaluation: *What is good about the partial sums concept?*

Synthesis: *How would you design a template for reordering partial sums?*

COMPLEX

Rigorous Problem Solving with the Concept of Partial Sums 1

After a long day of collecting seashells on the beach, Kayla separated the perfect seashells from the chipped seashells. There were 39 chipped seashells and 25 perfect seashells.

Part A

Which equation uses a place value strategy to represent the total number of seashells Kayla collected ?

- o $39 + 25 = 50 + 14$
- o $39 + 25 = 12 + 7$
- o $39 + 25 = 50 + 4$
- o $39 + 25 = 38 + 24$
- o $39 + 25 = 50 + 16$

Part B

What if Kayla found 3 more perfect seashells and 6 more chipped seashells?

Write an equation that uses a place value strategy to represent the new total of seashells Kayla collected.

Rigorous Problem Solving with the Concept of Partial Sums 2

After another long day of collecting seashells on the beach, Kayla separated the perfect seashells from the chipped seashells. There were 28 chipped seashells and 17 perfect seashells.

Part A

Which equation uses a place value strategy to represent the total number of seashells Kayla collected ?

o $28 + 17 = 30 + 16$
o $28 + 17 = 10 + 17$
o $28 + 17 = 30 + 7$
o $28 + 17 = 28 + 8$
o $28 + 17 = 30 + 15$

Part B

What if Kayla found 4 more perfect seashells and 5 more chipped seashells?

Write an equation that uses a place value strategy to represent the new total of seashells Kayla collected.

Race to the Top

A game designed to build the concept of partial sums

Materials:

- Race to the Top Game Board
- Four sets of 0–9 Number Cards
- Two Pawns

Directions:

1. Players shuffle each set of number cards and place cards face down in a pile near each of the template boxes on the game board.
2. Players decide the ladder each person will use and place a pawn at the bottom of this ladder.
3. The player on the left flips over the top card in the first pile and places it in the first box. He flips over the top card in the second pile and places it in the second box.
4. The player on the right does the same with the third and fourth piles of cards and boxes.
5. Players read the equation. The goal is not to name the sum. Rather, the goal is to name the partial sums. For example, 36 + 27 = 50 + 13.
6. The first player to name the correct partial sums moves his pawn one step up the ladder. Used cards are placed under each pile of cards.
7. Play continues until the winning player reaches the top of the ladder.

Content Differentiation:

Moving Back: Use base ten blocks or other manipulatives to show the partial sums.

Moving Ahead: Try this game with 3-digit numbers.

Race to the Top Game Board

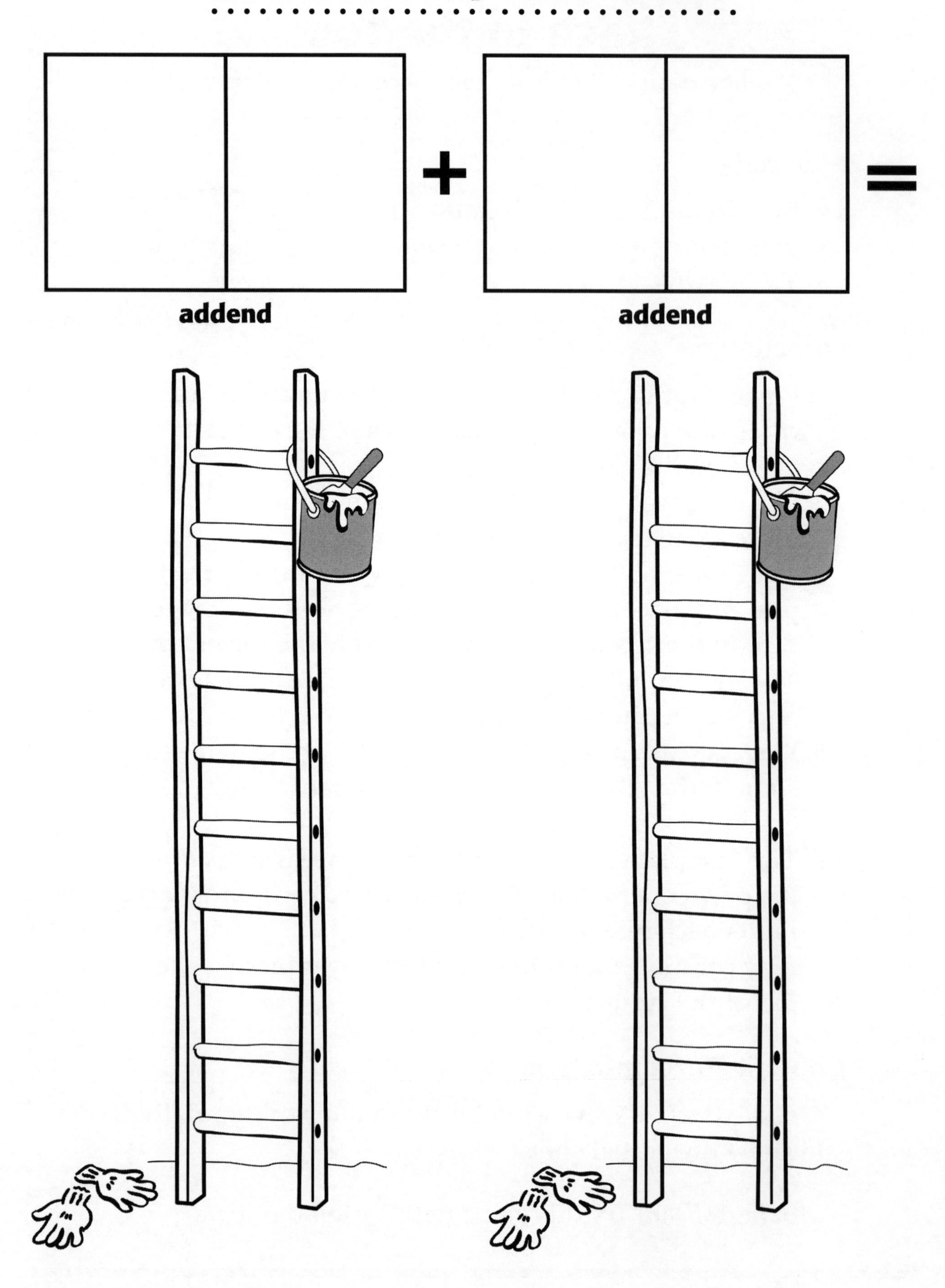

0–9 Number Cards

CONCEPT:

Partial Differences

What is the Partial Differences Concept?

The partial differences concept is a subtraction strategy that involves combining differences into workable clusters. Like partial sums, this concept is an alternative to the traditional algorithm. Instead of following rules that are meaningless to some students, students using the partial differences strategy do not have to *align columns, start with the ones,* or *cross out.* Instead, they find the difference between the tens, then find the difference between the ones, and combine these differences. For example, $57 - 26 = (50 - 20) + (7 - 6)$. Here, the difference between the tens is 30 and the difference between the ones is +1. These partial differences combine to $30 + 1 = 31$. The partial differences concept is also an effective strategy when a subtraction situation involves decomposing. For example, $84 - 36 = (80 - 30) + (4 - 6)$. Here, the difference between the tens is 50 and the difference between the ones is –2. These partial differences combine to $50 - 2 = 48$.

CCSS

Operations and Algebraic Thinking
Number and Operations in Base Ten

Formative Assessment

To find out if a student understands partial differences, ask the student to solve the following equations:

$37 - 12 =$ $45 - 23 =$ $57 - 34 =$
$64 - 25 =$ $52 - 36 =$ $70 - 28 =$

Ask the student to explain how she solves each problem. Invite the student to try the problems without aligning the digits in columns, without starting with the ones, and without crossing out.

Successful Strategies

Using number lines that include negative numbers helps students understand how to find the partial differences when the ones digit value of the minuend is greater than the ones digit value of the subtrahend. It also helps us focus on mathematical truths. It really is a lie when we tell students *you cannot take 6 from 4*. You can take 6 from 4. The result is –2. When the partial differences are combined, you just subtract without the need to memorize rules about adding positive and negative numbers. As with partial sums, even if students successfully use the traditional algorithm, learning the partial differences concept strengthens their conceptual understanding, increases mental math capacity, and improves flexibility and fluency.

Math Words to Use

partial, part, difference, subtract, subtraction, strategy, equation, equal, subtrahend, minuend

Questions at Different *Levels of Cognitive Demand*

EASY

Recall: *What is 46 – 21?*

Comprehension: *How did you solve the problem?*

Application: *What picture could you draw to show partial differences?*

Analysis: *How would you compare the partial differences?*

Evaluation: *What is helpful about the partial differences concept?*

Synthesis: *What if the partial differences were half as many?*

COMPLEX

Rigorous Problem Solving with the Concept of Partial Differences 1

Kolby is preparing for the upcoming football season. To increase his speed and agility he completes 84 ladder drills every hour of practice. Of the ladder drills completed each hour, 37 are lateral and the rest are forward. How many ladder drills each hour are forward?

Part A

Which equation uses a place value strategy to represent the total number of forward ladder drills Kolby completes each hour?

o $84 - 37 = 50 + 4$
o $84 - 37 = 50 - 4$
o $84 - 37 = 50 - 3$
o $84 - 37 = 50 + 3$
o $84 - 37 = 50 + 17$

Part B

What if Kolby completed 2 more lateral ladder drills per hour?

Write an equation that uses a place value strategy to represent the new number of forward ladder drills per hour.

Rigorous Problem Solving with the Concept of Partial Differences 2

Kolby is still preparing for the upcoming football season. To increase his speed and agility he completes 73 ladder drills every hour of practice. Of the ladder drills completed each hour, 45 are lateral and the rest are forward. How many ladder drills each hour are forward?

Part A

Which equation uses a place value strategy to represent the total number of forward ladder drills Kolby completes each hour?

o $73 - 45 = 30 + 5$
o $73 - 45 = 30 + 1$
o $73 - 45 = 30 - 1$
o $73 - 45 = 50 + 2$
o $73 - 45 = 30 - 2$

Part B

What if Kolby completed 4 more lateral ladder drills per hour?

Write an equation that uses a place value strategy to represent the new number of forward ladder drills per hour.

Race to the Bottom

A game designed to build the concept of partial differences

Materials:

- Two sets of 0–9 Number Cards *(see page 72)*
- One set of 5–9 Number Cards
- One set of 1–4 Number Cards
- Race to the Bottom Game Board
- Number Line and two pawns

Directions:

1. Players shuffle each set of number cards and place the cards face down in a pile near each of the template boxes on the game board. *Note: The 5–9 cards are used for the tens digit of the minuend. The 1–4 cards are used for the tens digit of the subtrahend.*
2. Players decide the ladder each person will use and place a pawn at the top of this ladder.
3. The player on the left flips over the top card in the first pile and places it in the first box. He flips over the top card in the second pile and places it in the second box.
4. The player on the right does the same with the third and fourth piles of cards and boxes.
5. Players read the equation. The goal is not to name the difference. Rather, the goal is to name the partial differences, for example, $63 - 27 = 40 - 4$. Use the number line as needed.
6. The first player to name the correct partial differences moves his pawn one step down the ladder. Used cards are placed under each pile of cards. Play continues until the winning player reaches the bottom of his ladder.

Content Differentiation:

Moving Back: Use base ten blocks or other manipulatives to show the partial differences.

Moving Ahead: Try this game with 3-digit numbers.

Race to the Bottom Game Board

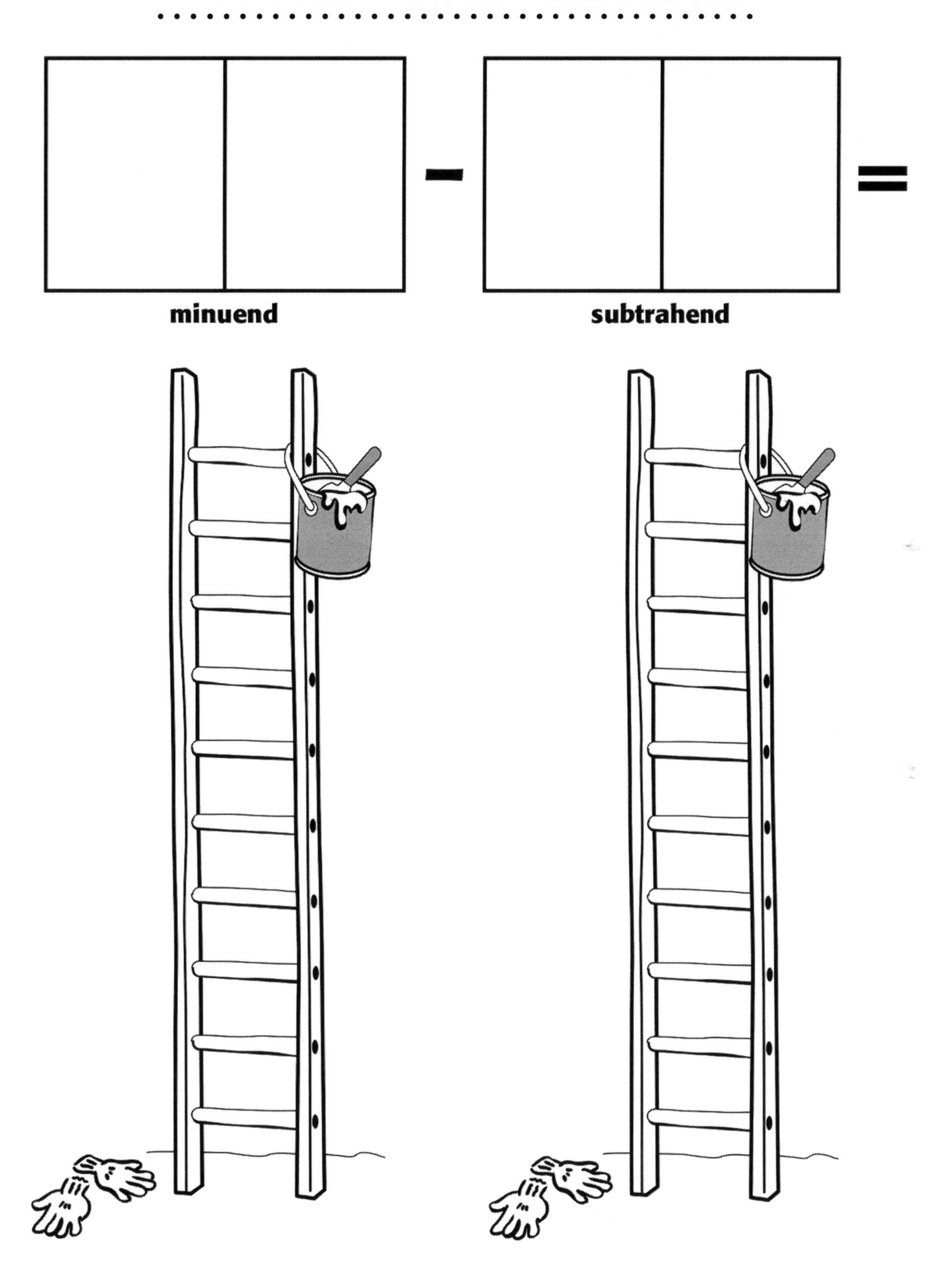

Number Line

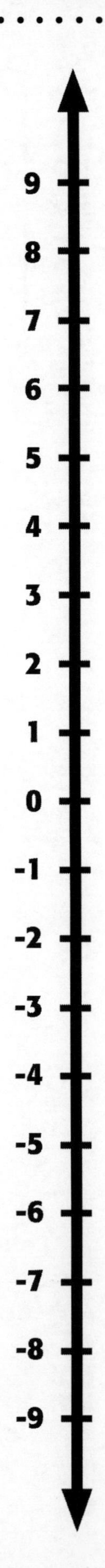

CONCEPT:

Near Tens for Addition and Subtraction

What is the Near Tens Concept?

The near tens concept involves adjusting numbers based on their proximity to ten. The near tens concept can be used in addition, for example, $57 + 9 = 57 + 10 - 1$ or $57 + 8 = 57 + 10 - 2$. The near tens concept can also be used in subtraction, for example $83 - 9 = 83 - 10 + 1$ or $83 - 8 = 83 - 10 + 2$. When students use the near tens concept, they apply the kind of flexibility with numbers that is needed for mental math and deep conceptual understanding.

CCSS

Operations and Algebraic Thinking
Number and Operations in Base Ten

Formative Assessment

To find out if a student understands the near tens concept, ask the student to solve the following equations;

$48 + 10 =$ $56 + 20 =$
$73 - 10 =$ $67 - 30 =$
$53 + 9 =$ $47 + 29 =$
$64 - 9 =$ $93 - 49 =$
$26 + 8 =$ $44 + 28 =$
$56 - 8 =$ $77 - 38 =$

Ask students to explain how they solve each problem. If the student uses the traditional algorithm (successfully or not successfully), ask the student if she knows another way to solve the equations. Invite the student to try the problems using what they know about adding and subtracting tens.

Successful Strategies

To be successful with the near tens concept, it is essential that students have ownership of plus or minus ten. The near tens concept builds upon what students know about adding and subtracting tens. For many students, using the hundred chart helps them to see and apply the near tens concept.

Math Words to Use

near ten, add, addition, subtract, subtraction, adjust, compensate

Questions at Different *Levels of Cognitive Demand*

EASY

Recall: *What is 97 – 10 ?*

Comprehension: *How could you use a near ten to solve 43 + 29?*

Application: *How would you show the near tens concept with manipulatives?*

Analysis: *Which problems work better with the near tens concept?*

Evaluation: *When is it not a good plan to use the near tens concept?*

Synthesis: *How would you teach the near tens concept to someone who did not know it?*

COMPLEX

Rigorous Problem Solving with the Concept of Near Tens for Addition and Subtraction 1

The students at Blaise Elementary School shop in the school store. Below is a chart of the sales during the first week.

Item	Started with	Sold	How many left
Gel pens	56	39	
Erasers	83	59	
Notebooks	34	29	

Part A

Which equation uses a place value strategy to represent how many gel pens are left in the school store?

o $56 - 39 = 56 - 30 + 1$
o $56 - 39 = 56 - 30 - 1$
o $56 - 39 = 56 - 40 + 1$
o $56 - 39 = 56 - 40 - 1$
o $56 - 39 = 56 - 50 + 1$

Part B

Use a place value strategy to represent how many erasers and notebooks are left in the school store.

Erasers

Notebooks

Rigorous Problem Solving with the Concept of Near Tens for Addition and Subtraction 2

The students at Blaise Elementary School shop in the school store. Below is a chart of the sales during the second week.

Item	Started with	Sold	How many left
Gel pens	73	49	
Erasers	92	69	
Notebooks	58	39	

Part A

Which equation uses a place value strategy to represent how many gel pens are left in the school store?

o $73 - 49 = 73 - 40 + 1$
o $73 - 49 = 73 - 40 - 1$
o $73 - 49 = 73 - 30 + 1$
o $73 - 49 = 73 - 50 - 1$
o $73 - 49 = 73 - 50 + 1$

Part B

Use a place value strategy to represent how many erasers and notebooks are left in the school store.

Erasers	Notebooks

Drop It

A game designed to build the concept of near tens

Materials:

- Die (labeled 18, 28, 39, 49, 40, 50)
- Plus/Minus Spinner, paper clip and pencil
- Drop It Hundred Chart
- Chip or counter

Directions:

1. The first player spins the spinner to find out if he will add or subtract.
2. The first player drops the chip/counter on the hundred chart to find one of the numbers in his equation and rolls the die to find the other number in the equation. *Note: The chip/counter is dropped on the 1–50 section (shaded) for addition and the 51–100 section (not shaded) for subtraction. This number serves as the first addend or the minuend. The number on the die serves as the second addend or the subtrahend.* For example, 74 – 39 = ? The player touches 74 on the hundred chart, moves his finger up four rows of ten (– 40) and forward one space to the right (+ 1). The player describes the near tens situation, for example, 74 – 39 = 74 – 40 + 1.
3. The sum/difference is the player's positive score for the round.
4. The next player spins the spinner, drops the chip/counter, rolls the die, models, and describes the near tens equation.
5. Players add their scores from each round. The winning player is the player with the highest score after 10 rounds.

Content Differentiation:

Moving Back: Try the game with a die labeled 8, 8, 9, 9, 10, 10.

Moving Ahead: Try the game with a die labeled 38, 48, 58, 69, 79, 89 using a two-hundred chart.

Plus/Minus Spinner

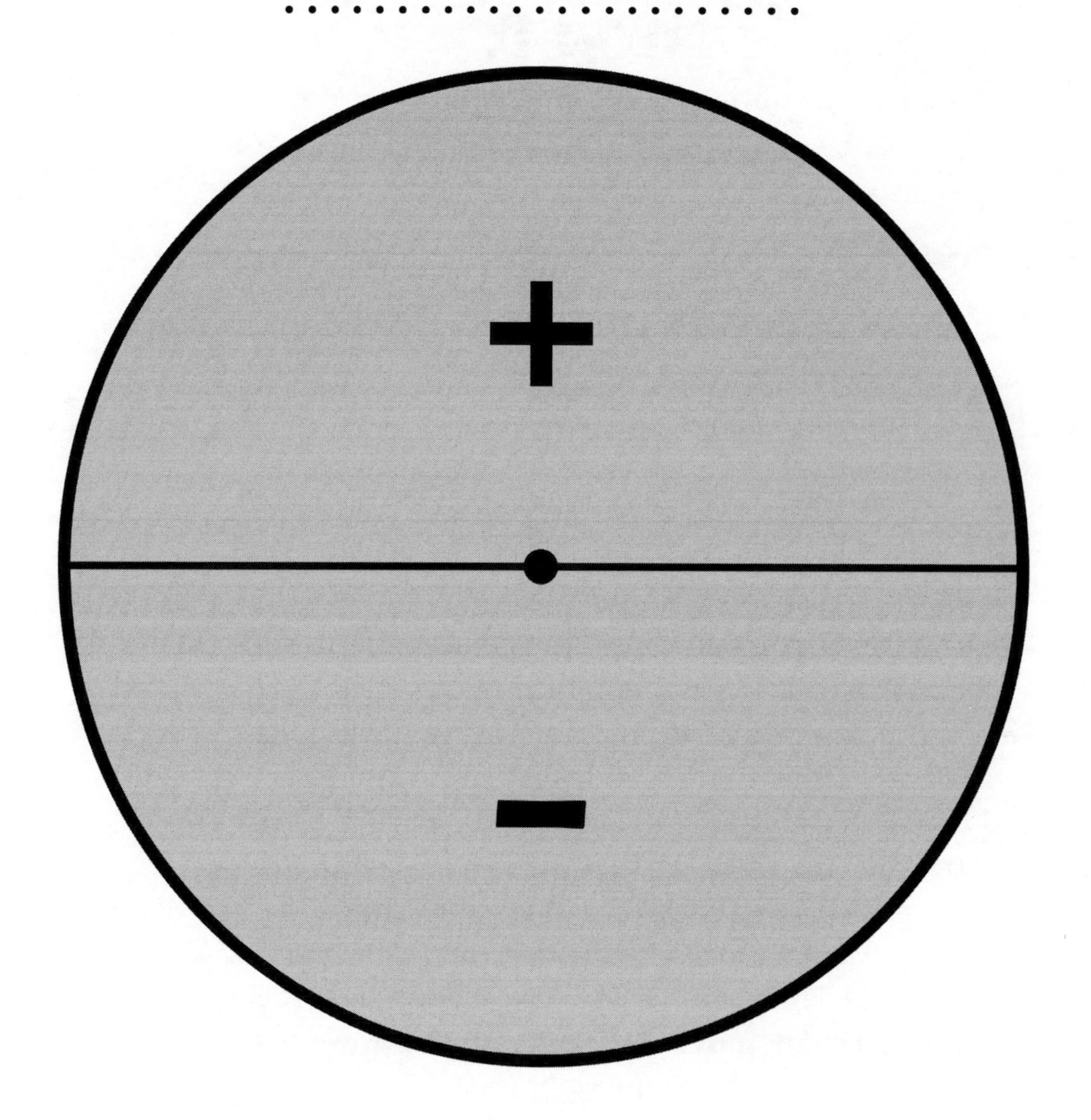

To use the spinner, place a paper clip in the center. Place pencil point through the paper clip at the center of the spinner. Holding the pencil securely with one hand, spin the paper clip with the other hand.

Drop It Hundred Chart

1	2	3	4	5	6	7	8	9	10
11	12	13	14	15	16	17	18	19	20
21	22	23	24	25	26	27	28	29	30
31	32	33	34	35	36	37	38	39	40
41	42	43	44	45	46	47	48	49	50
51	52	53	54	55	56	57	58	59	60
61	62	63	64	65	66	67	68	69	70
71	72	73	74	75	76	77	78	79	80
81	82	83	84	85	86	87	88	89	90
91	92	93	94	95	96	97	98	99	100

CONCEPT:

Equal Differences

What is the Equal Difference Concept?

The equal difference concept is a subtraction strategy based on the idea that different subtraction expressions can have the same difference. The focus is on the value between numbers, for example, $76 - 29 = 77 - 30$. By adding one to the subtrahend and one to the minuend, we create an expression that is more easily solved. In some situations, it makes sense to add two to the subtrahend and the minuend, for example, $52 - 28 = 54 - 30$. In some situations, we may subtract from the subtrahend and the minuend to create a new expression with the same difference. For example, $68 - 41 = 67 - 40$ or $96 - 82 = 94 - 80$. Understanding how to mentally create expressions with the same differences helps students learn to build flexibility and fluency with numbers and operations.

CCSS

Operations and Algebraic Thinking
Number and Operations in Base Ten

Formative Assessment

To find out if a student understands the equal difference concept, ask the student if the following equations are true or not:

$54 - 29 = 55 - 30$	$87 - 34 = 90 - 31$
$76 - 48 = 78 - 50$	$81 - 34 = 80 - 33$

Ask the student to explain why these equations are true or not true (the third equation is NOT true). Notice if the student solves each side of the equation to verify the answer or if the student uses the equal difference concept.

Successful Strategies

The equal difference concept is applicable in any given subtraction equation, but it is most efficient when the created expression invites mental math. For example, 73 – 25 = 74 – 26. The created expression does not necessarily increase opportunities for mental math. It is beneficial for students to think about when to use the equal difference concept and when it may be better to use a different concept.

Math Words to Use

equal, difference, subtract, subtraction, add, addition, expression, equation

Questions at Different *Levels of Cognitive Demand*

Recall: *What is the difference between 52 and 39?*

Comprehension: *Which examples could you give of the equal difference concept?*

Application: *How would you show the equal difference concept on a number line?*

Analysis: *How are these expressions related?*

Evaluation: *Which problems work best with the equal difference concept?*

Synthesis: *What if you were to create a new way to record twice the difference?*

COMPLEX

Rigorous Problem Solving with the Concept of Equal Differences 1

Kayla announced that $93 - 58 = 95 - 60$. Do you agree or disagree?

Part A

Is Kayla correct or incorrect? ___________ Use the number line to prove you answer.

50 51 52 53 54 55 56 57 58 59 60 61 62 63 64 65 66 67 68 69 70 71 72 73 74 75 76 77 78 79 80 81 82 83 84 85 86 87 88 89 90 91 92 93 94 95 96 97 98 99 100

Part B

If 62 pennies were tossed in air and 39 of them landed on heads, how many landed on tails?

Use a place value strategy to prove your answer.

Part C

If 73 pennies were tossed in air and 41 of them landed on heads, how many landed on tails?

Use a place value strategy to prove your answer.

Rigorous Problem Solving with the Concept of Equal Differences 2

Joey announced that 84 – 68 = 86 – 70. Do you agree or disagree?

Part A

Is Joey correct or incorrect? ____________ Use the number line to prove you answer.

50 51 52 53 54 55 56 57 58 59 60 61 62 63 64 65 66 67 68 69 70 71 72 73 74 75 76 77 78 79 80 81 82 83 84 85 86 87 88 89 90 91 92 93 94 95 96 97 98 99 100

Part B

If 86 pennies were tossed in air and 39 of them landed on heads, how many landed on tails?

Use a place value strategy to prove your answer.

Part C

If 65 pennies were tossed in air and 31 of them landed on heads, how many landed on tails?

Use a place value strategy to prove your answer.

All Strung Out

A game designed to build the concept of equal difference

Materials:

- **Number Cards (50–98)**
- **One Subtrahend Spinner (+2 or +1 or –1)**
- **Paper clip and pencil**
- **Number line (1–100)** *Note: If number line is not available, cut apart the rows of a hundred chart and tape the ends together.*
- **String (or ribbon or strips of paper), scissors**

Directions:

1. **Shuffle the number cards and place face-down in a pile. The first player draws a card. This number will be the minuend.**
2. **The first player spins the spinner to find the subtrahend for the subtraction expression. This player cuts a piece of string that shows the difference (distance between the minuend and the subtrahend). The player moves the string two numbers to the right (when using the +2 spinner) and describes the equal difference equation using the term "is the same as." For example, 86 – 28 *is the same as* 88 – 30.**
3. **If the player accurately models and describes the equal difference equation, the player keeps the string.**
4. **Players take turns drawing number cards, spinning, modeling, and describing the equal difference expressions.**
5. **After each player has five turns, all players line up their pieces of string, and compare their total lengths. The winner is the player with the most string.**

Content Differentiation:

Moving Back: **Use the +1 spinner (move string one number to the right).**
Moving Ahead: **Use the –1 spinner (move string one number to the left) or make your own spinner.**

All Strung Out Number Cards

50	51	52	53	54
55	56	57	58	59
60	61	62	63	64
65	66	67	68	69
70	71	72	73	74

All Strung Out Number Cards

75	76	77	78	79
80	81	82	83	84
85	86	87	88	89
90	91	92	93	94
95	96	97	98	

Subtrahend Spinner Options

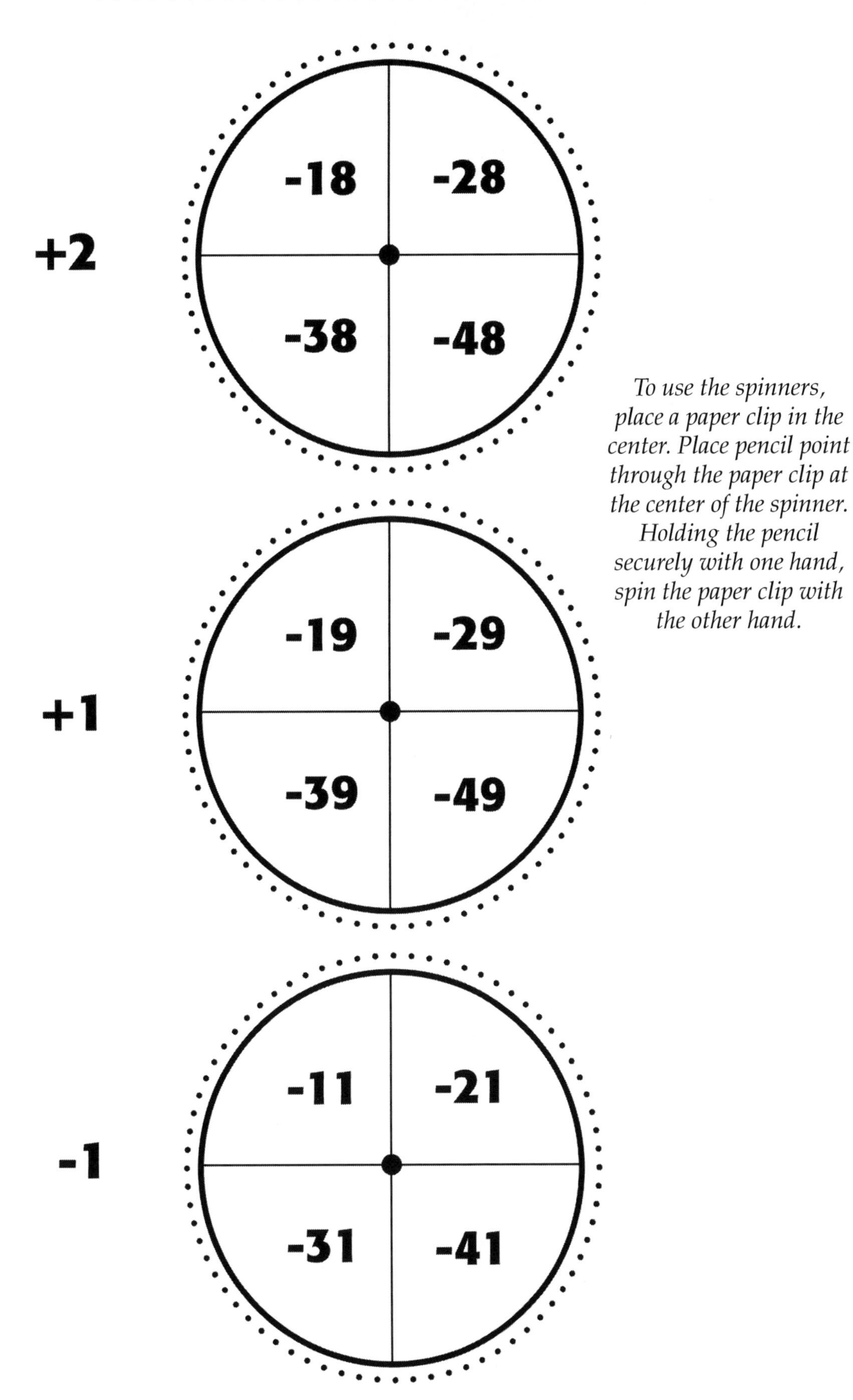

To use the spinners, place a paper clip in the center. Place pencil point through the paper clip at the center of the spinner. Holding the pencil securely with one hand, spin the paper clip with the other hand.

CHAPTER

Multiplication and Division Concepts

Repeated Addition Multiplication

Repeated Subtraction Division

Fair Shares Division

Rectangular Arrays for Multiplication and Division

Multiplying and Dividing by Ten

Perfect Squares and Near Squares

Near Tens for Multiplication

Fact Families—Multiplication and Division

Partial Products

Partial Quotients

Equal Products

CONCEPT:

Repeated Addition Multiplication

What is the Repeated Addition Multiplication Concept?

When we multiply, the result is the total number, referred to as the product. In multiplication, we combine a certain number of groups, which is the multiplier. Each group has the same quantity, which is the multiplicand. Because addition and multiplication are interrelated operations, we obtain the same results by repeated addition. For example, 3 x 5 = 5 + 5 + 5. There are 3 groups of 5. Some students are encouraged to memorize multiplication facts before they understand the concept of multiplication. The result is inconsistent surface knowledge. Working with repeated addition builds the necessary foundations for permanent ownership of multiplication facts.

CCSS

Operations and Algebraic Thinking

Formative Assessment

To find out if a student understands repeated addition multiplication, give the student the following repeated addition equations and ask the student to give the related multiplication equation.

3 + 3 + 3 + 3 =   8 + 8 + 8 =

If the student is successful, provide the following multiplication equations and ask the student to give the related repeated addition equations.

7 x 4 =   6 x 9 =

If the student is not successful with either set of problems, try groups of two, five, or ten. If the student is successful with both sets of problems, try more complex problems.

Successful Strategies

When working with the repeated addition multiplication concept, it is often beneficial to model the groups with manipulatives or show the groups with pictures. It is also helpful to put the mathematical situations in context. For example, 4 + 4 + 4 can be placed in the context of three cars with four people in each car. Students can draw pictures of the cars and people or use manipulatives to show the groups of people. In addition, it is beneficial to begin making the connection to multiplication by using the phrase, "groups of." Saying, "There are 3 groups of 4," is a helpful bridge from 4 + 4 + 4 to 3 x 4. The word "times" may cause confusion for students. Beginning with "groups of" helps students to understand the concept and progress to other ways to communicate the concept such as using the word "times."

Math Words to Use

repeat, repeated, addition, add, multiply, multiplication, groups, sets, multiplier, multiplicand, product

Questions at Different *Levels of Cognitive Demand*

Recall: *What is 2 + 2 + 2 + 2 + 2?*

Comprehension: *How do know that 5 + 5 + 5 + 5 = 20?*

Application: *How would you illustrate 6 groups of 3?*

Analysis: *How does skip counting connect to repeated addition?*

Evaluation: *Why is repeated addition a multiplication concept?*

Synthesis: *What kind of diagram could you create to show repeated addition?*

Rigorous Problem Solving with the Concept of Repeated Addition Multiplication 1

Five third graders will celebrate their birthdays at a party on Friday. Each student will receive one cake. Each cake needs eight candles on it.

Part A

Which expressions represent the number of candles needed?

o $3 + 5$
o 3×5
o $3 + 1 + 8$
o $5 + 5 + 5$
o $8 + 8 + 8 + 8 + 8$
o $8 + 8 + 8$
o 3×8
o 5×8
o 1×8
o $1 + 8$

Part B

If each cake is cut into 6 pieces, how many pieces of cake will there be?

Justify your answers with an equation.

Rigorous Problem Solving with the Concept of Repeated Addition Multiplication 2

Five fourth graders will celebrate their birthdays at a party on Friday. Each student will receive one cake. Each cake needs nine candles on it.

Part A

Which expressions represent the number of candles needed?

o $4 + 5$
o 4×5
o $5 + 1 + 9$
o $5 + 5 + 5 + 5$
o $9 + 9 + 9 + 9 + 9$
o $9 + 9 + 9$
o 4×9
o 5×9
o 1×9
o $1 + 9$

Part B

If each cake is cut into 8 pieces, how many pieces of cake will there be?

Justify your answers with an equation.

Cakes and Candles

A game designed to build the concept of repeated addition multiplication

Materials:

- One white die (labeled 1–6 with numerals)
- One red die (labeled 1–6 with numerals)
- Paper and pencil (or dry erase board and marker) for each player

Directions:

1. The first player rolls the dice. The white die represents the cakes. The red die represents the candles on each cake.
2. Players race to draw the number of cakes and the number of candles on each cake. For example, if the number on the white die is 3 and the number on the red die is 2, the illustration may look like this . . .

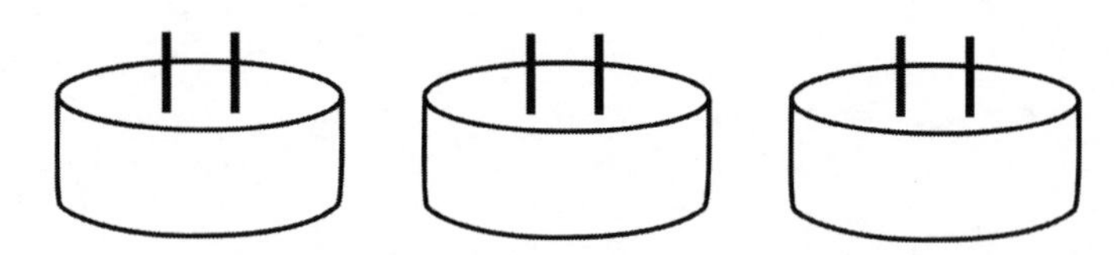

3. The first player to complete the picture and call out the total number of candles (product) scores two points.
4. If the player who rolled the dice can state the "groups of" equation, she earns one point. For example, "3 groups of 2 equal 6."
5. Players continue taking turns rolling, illustrating, calling out the product, and giving equations.
6. The winner is the first player to score 20 or greater.

Content Differentiation:

Moving Back: Use dice labeled 1, 1, 2, 2, 3, 3 with numerals.

Moving Ahead: Use decahedron dice (labeled 0–9) or 0–9 spinner *(see page 142)*.

CONCEPT:

Repeated Subtraction Division

What is the Repeated Subtraction Division Concept?

When we divide, we find the number of times a number is divisible; the result is the quotient. The number that is divided is called the dividend. The number that divides the dividend is called the divisor. Because subtraction and division are interrelated operations, we obtain the same results by repeated subtraction. For example, $15 \div 3$ can be solved by subtracting. Groups of 3 are subtracted from the dividend until zero is reached. We keep track of how many groups are subtracted to find the quotient. Some students are encouraged to memorize division facts before they understand the concept of division. Working with repeated subtraction as a way to understand division builds necessary foundations for students.

CCSS

Operations and Algebraic Thinking

Formative Assessment

To find out if a student understands repeated subtraction division, give him the following repeated subtraction equations and ask him to give the related division equation.

$16 - 4 - 4 - 4 - 4 =$ $\qquad$ $21 - 7 - 7 - 7 =$

If the student is successful, provide the following division equations and ask the student to give the related repeated subtraction equations.

$35 \div 5 =$ $\qquad$ $56 \div 8 =$

If the student is not successful with either set of problems, try subtracting groups of two, five, or ten. If he is successful with both sets of problems, try more complex problems.

Successful Strategies

When working with the repeated subtraction division concept, it is often beneficial to model the subtraction of groups with manipulatives or to show the subtraction of groups with pictures. It is also helpful to put the mathematical situations in context. For example, 12 – 4 – 4 – 4 can be placed in the context of twelve cookies placed in sets of four on plates. Students can draw pictures of the plates and cookies or use manipulatives to show the sets of cookies. It is also beneficial to begin making the connection to division by continually asking, "How many groups have been subtracted?" and "Are there enough left to subtract another group?" These questions help strengthen students' understanding of division.

Math Words to Use

repeat, repeated, subtraction, subtract, divide, division, groups, sets, dividend, divisor, quotient

Questions at Different *Levels of Cognitive Demand*

Recall: *What is 8 – 2 – 2 – 2 – 2?*

Comprehension: *How do know that 15 – 5 – 5 – 5 = 0?*

Application: *How could you use repeated subtraction to solve 20 ÷ 4?*

Analysis: *How would you compare this repeated subtraction problem with another?*

Evaluation: *What would you recommend to help someone learn repeated subtraction?*

Synthesis: *What if there are leftovers after the repeated subtraction?*

COMPLEX

Rigorous Problem Solving with the Concept of Repeated Subtraction Division 1

The friends want to equally share 24 cookies.

Part A

Which equations represent the friends' equally sharing 24 cookies?

- o $24 \div 8 = 3$
- o $24 - 6 - 6 - 6 - 6 = 0$
- o $24 - 4 - 5 - 5 - 5 - 5 = 0$
- o $24 \div 4 = 8$
- o $24 - 12 - 12 = 0$
- o $24 \div 6 = 12$

Part B

What if the friends want to equally share 20 cookies?

How many friends? [ ]

How many cookies for each friend? [ ]

Write the equation. [ ]

Write an equation that represents how to equally share 20 cookies a different way?

[ ]

Rigorous Problem Solving with the Concept of Repeated Subtraction Division 2

The friends want to equally share 30 cookies.

Part A

Which equations represent the friends' equally sharing 30 cookies?

- o $30 \div 3 = 10$
- o $30 - 15 - 15 = 0$
- o $30 \div 6 = 15$
- o $30 \div 10 = 3$
- o $30 - 6 - 6 - 6 - 6 - 6 = 0$
- o $30 - 6 - 5 - 5 - 5 - 5 = 0$

Part B

What if the friends want to equally share 18 cookies?

How many friends? [ ]

How many cookies for each friend? [ ]

Write the equation. [ ]

Write an equation that represents how to equally share 18 cookies a different way?

[ ]

Cookies for My Friends

A game designed to build the concept of repeated subtraction division

Materials:

- 10 paper plates
- 100 counting chips
- Cookie Cards cut apart and placed in a cup
- Paper and pencil (or dry erase board and marker) for each player

Directions:

1. The first player chooses a random cookie card from the cup. The player gathers this many counting chips to represent cookies.
2. The player reads how many cookies each friend gets and begins subtracting out sets of cookies represented by chips. The player places sets of cookies on plates. Other players consider how many friends will share the cookies/how many plates will be used/how many times the divisor can be subtracted. The other players race to draw a picture of what the model will look like when complete.
3. The first person who completes an accurate illustration scores one point. *Note: The player modeling the repeated subtraction with chips and plates does not try to score.*
4. Players take turns choosing cookie cards, modeling different repeated subtraction scenarios, and illustrating the models.
5. The winner is the first player to earn ten points.

Content Differentiation:

Moving Back: Use first five rows of cookie cards.

Moving Ahead: Make your own cookie cards. Try remainders.

Cookie Cards

1 cookie Each friend gets 1 cookie	**2 cookies** Each friend gets 1 cookie	**3 cookies** Each friend gets 3 cookies	**4 cookies** Each friend gets 1 cookie	**5 cookies** Each friend gets 5 cookies
6 cookies Each friend gets 1 cookie	**7 cookies** Each friend gets 7 cookies	**8 cookies** Each friend gets 1 cookie	**9 cookies** Each friend gets 9 cookies	**10 cookies** Each friend gets 1 cookie
4 cookies Each friend gets 2 cookies	**6 cookies** Each friend gets 3 cookies	**8 cookies** Each friend gets 2 cookies	**10 cookies** Each friend gets 5 cookies	**12 cookies** Each friend gets 6 cookies
14 cookies Each friend gets 2 cookies	**16 cookies** Each friend gets 8 cookies	**18 cookies** Each friend gets 2 cookies	**20 cookies** Each friend gets 10 cookies	**9 cookies** Each friend gets 3 cookies
12 cookies Each friend gets 3 cookies	**15 cookies** Each friend gets 5 cookies	**18 cookies** Each friend gets 3 cookies	**21 cookies** Each friend gets 7 cookies	**24 cookies** Each friend gets 3 cookies
27 cookies Each friend gets 9 cookies	**30 cookies** Each friend gets 3 cookies	**16 cookies** Each friend gets 4 cookies	**20 cookies** Each friend gets 4 cookies	**24 cookies** Each friend gets 6 cookies
28 cookies Each friend gets 4 cookies	**32 cookies** Each friend gets 8 cookies	**36 cookies** Each friend gets 4 cookies	**40 cookies** Each friend gets 10 cookies	**25 cookies** Each friend gets 5 cookies
30 cookies Each friend gets 5 cookies	**35 cookies** Each friend gets 7 cookies	**40 cookies** Each friend gets 5 cookies	**45 cookies** Each friend gets 9 cookies	**50 cookies** Each friend gets 5 cookies
36 cookies Each friend gets 6 cookies	**42 cookies** Each friend gets 7 cookies	**48 cookies** Each friend gets 6 cookies	**54 cookies** Each friend gets 9 cookies	**60 cookies** Each friend gets 6 cookies
49 cookies Each friend gets 7 cookies	**56 cookies** Each friend gets 8 cookies	**63 cookies** Each friend gets 7 cookies	**70 cookies** Each friend gets 10 cookies	**64 cookies** Each friend gets 8 cookies
72 cookies Each friend gets 8 cookies	**80 cookies** Each friend gets 10 cookies	**81 cookies** Each friend gets 9 cookies	**90 cookies** Each friend gets 9 cookies	**100 cookies** Each friend gets 10 cookies

CONCEPT:

Fair Shares Division

What is the Fair Shares Division Concept?

The fair shares concept is similar to the repeated subtraction concept because it is division. The fair shares concept is different from the repeated subtraction concept because the number of groups is known but the amount in each group is unknown. With repeated subtraction the number in each group is known but the number of groups is unknown. When students work with fair shares they are partitioning or splitting up the dividend into a specific number of groups. When students partition a given quantity, they distribute equal amounts, one at a time, much like when a person deals out cards.

CCSS

Operations and Algebraic Thinking

Formative Assessment

To find out if a student understands fair shares division, give the student a set of 48 counters and explain that the student will use counters to model the problem.

There are 48 pieces of candy. There are four boxes. How many pieces of candy go inside each box?

Notice if and how the student partitions the dividend. If the student is not successful, try a similar story with easier numbers. If the student is successful try a similar story using numbers that result in a remainder situation.

Successful Strategies

Students benefit from engaging in fair shares division by actually partitioning, not just watching someone partition. This active learning helps students to build background experiences and solidify their learning. Likewise, an illustration of fair shares division should not replace actual experiences. However, an illustration following the partitioning experience serves as a beneficial way to communicate the mathematics. To make the division connections while students are partitioning teachers should ask, "How many are in each group so far?" and "How many more do you think each group will get?" These questions and others help students strengthen their understanding of division.

Math Words to Use

fair shares, groups, sets, remainder, divide, division, dividend, divisor, quotient

Questions at Different *Levels of Cognitive Demand*

Recall: *What is the divisor?*

Comprehension: *How would you describe fair shares division?*

Application: *When would you use fair shares to solve a problem?*

Analysis: *How would you compare fair shares to repeated subtraction?*

Evaluation: *In your judgement, is the fair shares concept helpful? Why?*

Synthesis: *What kind of song could you create about fair shares division?*

COMPLEX

Rigorous Problem Solving with the Concept of Fair Shares Division 1

Kass Marie has 50 pennies. How many banks does she need to represent fair share division of the 50 pennies? How many pennies will be in each bank?

Part A

Complete the chart to represent different ways to evenly divide 50 pennies.

Piggy Banks	Pennies in each Bank	Equation
2		$50 \div 2 =$
	10	
10		
	2	

Part B

Kayla Blaise thinks 60 pennies can be equally divided into 5 piggy banks.

Do you agree or disagree?

Explain your answer.

Rigorous Problem Solving with the Concept of Fair Shares Division 2

Kolby has 40 pennies. How many banks does she need to represent fair share division of the 40 pennies? How many pennies will be in each bank?

Part A

Complete the chart to represent different ways to evenly divide 40 pennies.

Piggy Banks	Pennies in each Bank	Equation
2		$40 \div 2 =$
	10	
10		
	2	
	8	
8		

Part B

Kirkland thinks 42 pennies can be equally divided into 6 piggy banks.

Do you agree or disagree? [ ]

Explain your answer.

[ ]

Piggy Banks

A game designed to build the concept of fair shares division

PLAYERS: 2 OR MORE

Materials:

- 100 pennies
- Hundred chart
- Scissors
- Cup
- 12 Piggy Banks

Directions:

1. Cut apart the numbers in the hundred chart. Place the numbers 1–100 in a cup.
2. The first player chooses one random number from the cup. This number is used as the dividend. The player gathers pennies to represent the dividend. The player then decides how many piggy banks to use to try to have fair shares. The player lays out the piggy banks.
3. The player then distributes the pennies in equal amounts, one at a time to each bank.
4. If the player distributes fair shares, meaning there are no left over pennies, the player earns 1 point for each piggy bank used. If there is a remainder, no points are earned and players decide if a different amount of piggy banks could have been used.
5. Players take turns choosing dividends, deciding how many piggy banks, and distributing pennies.
6. The winner is the first player to score 50 points or greater.

Content Differentiation:

Moving Back: Use only the numbers that are multiples of 2, 5, or 10.

Moving Ahead: Play this game with the goal of having a remainder and/or use more piggy banks.

Piggy Banks

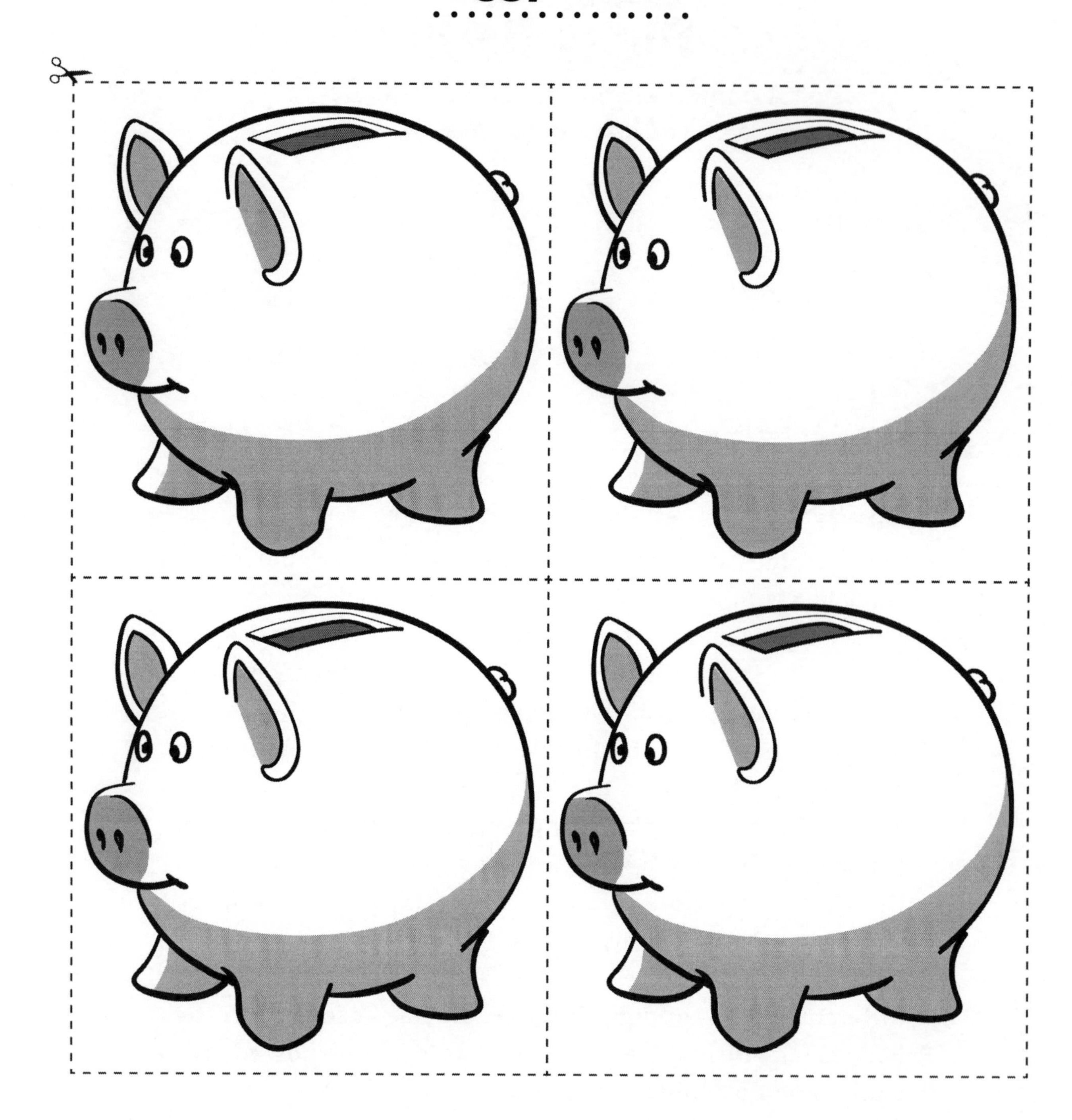

CONCEPT:

Rectangular Arrays for Multiplication and Division

What are Rectangular Arrays for Multiplication and Division?

Beginning with single-digits, students can model multiplication and division with rectangular arrays. To construct a rectangular array, students form or draw equal rows. For example, 4 x 8 or 32 ÷ 8 is shown as:

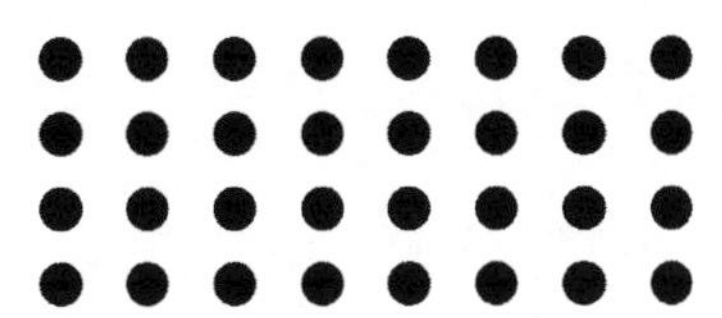

Initially, students may count by ones to find the product/ dividend. Building on this knowledge, we can encourage students to skip count and recognize the array without having to count all. After students gain comfort with single-digits, they can use rectangular arrays to solve multiplication and division problems with multi-digits.

CCSS

Operations and Algebraic Thinking

Formative Assessment

To find out if a student understands rectangular arrays for multiplication and division, ask the student to draw a rectangular array for the following expressions:

3 x 10	21÷7
7 x 4	40÷8

Notice how the student illustrates the arrays. Provide easier or more difficult expressions based on the student's degree of success with these problems.

Successful Strategies

Some teachers become too focused on rows versus columns when teaching rectangular arrays. By most definitions, rows are horizontal and columns are vertical. However, if you turn the array 90° the rows and columns switch. What complicates the issue further is that some educators use the term "rows" to mean both columns and rows. To deal with these complications, explain the inconsistencies to students and let them decide how they would like to use the terms in their descriptions of arrays. In this way, students can take ownership and focus on building meaningful experiences with multiplication and division.

Math Words to Use
rectangular array, rows, columns, multiply, multiplication, multiplier, multiplicand, product divide, division, dividend, divisor, quotient

Questions at Different *Levels of Cognitive Demand*

Recall: *What are three rows of two?*

Comprehension: *What are rectangular arrays?*

Application: *How would you show the array for 9 x 4?*

Analysis: *Which array matches the expression?*

Evaluation: *What is the most important thing about rectangular arrays?*

Synthesis: *How would you recognize the array to show a different problem with the same product/dividend?*

COMPLEX

Rigorous Problem Solving with the Concept of Rectangular Arrays for Multiplication and Division 1

The Allen Pond Park has the same number of frogs on each log.

Part A

The following array represents how many frogs on how many logs?

● ● ●
● ● ●
● ● ●
● ● ●
● ● ●
● ● ●

Logs ☐ Frogs on each log ☐ Total number of frogs ☐

Equation ☐

Part B

Draw a rectangular array that is twice as large as 6 x 2.

Total number of frogs ☐

Rigorous Problem Solving with the Concept of Rectangular Arrays for Multiplication and Division 2

The Allen Pond Park has the same number of frogs on each log.

Part A

The following array represents how many frog on how many logs?

● ● ● ●
● ● ● ●
● ● ● ●
● ● ● ●
● ● ● ●

Logs ☐ Frogs on each log ☐ Total number of frogs ☐

Equation ☐

Part B

Draw a rectangular array that is twice as large as 7 x 3.

Total number of frogs ☐

☐

Frogs on a Log

A game designed to build the concept of rectangular arrays for multiplication and division

Materials:

- 100 small plastic frogs (or centimeter cubes to represent frogs)
- 10 craft sticks (popsicle sticks)
- One 1–10 Spinner copied on white paper, paper clip
- One 1–10 Spinner copied on green paper, paper clip
- Dry erase boards and markers (or paper and pencil) for each player

Directions:

1. The first player spins the white spinner to determine the number of logs. The player displays this number with craft sticks lined up side-by-side.
2. The same player spins the green spinner to determine the number of frogs on each log. The player displays the frogs on the logs.
3. The other players illustrate the rectangular array using squares or circles on dry erase boards. The players also write the multiplication equation: [white] x [green] = [product]
4. The first player with the correct array and equation scores 100 points.
5. Players take turns modeling frogs on logs, drawing, and writing equations.
6. The winner is the first to score 1,000 or greater.

Content Differentiation:

Moving Back: Use white and green dice labeled 1–6 with numerals.

Moving Ahead: Play this game by writing division equations or both multiplication and division equations with or without the models.

1–10 Spinner

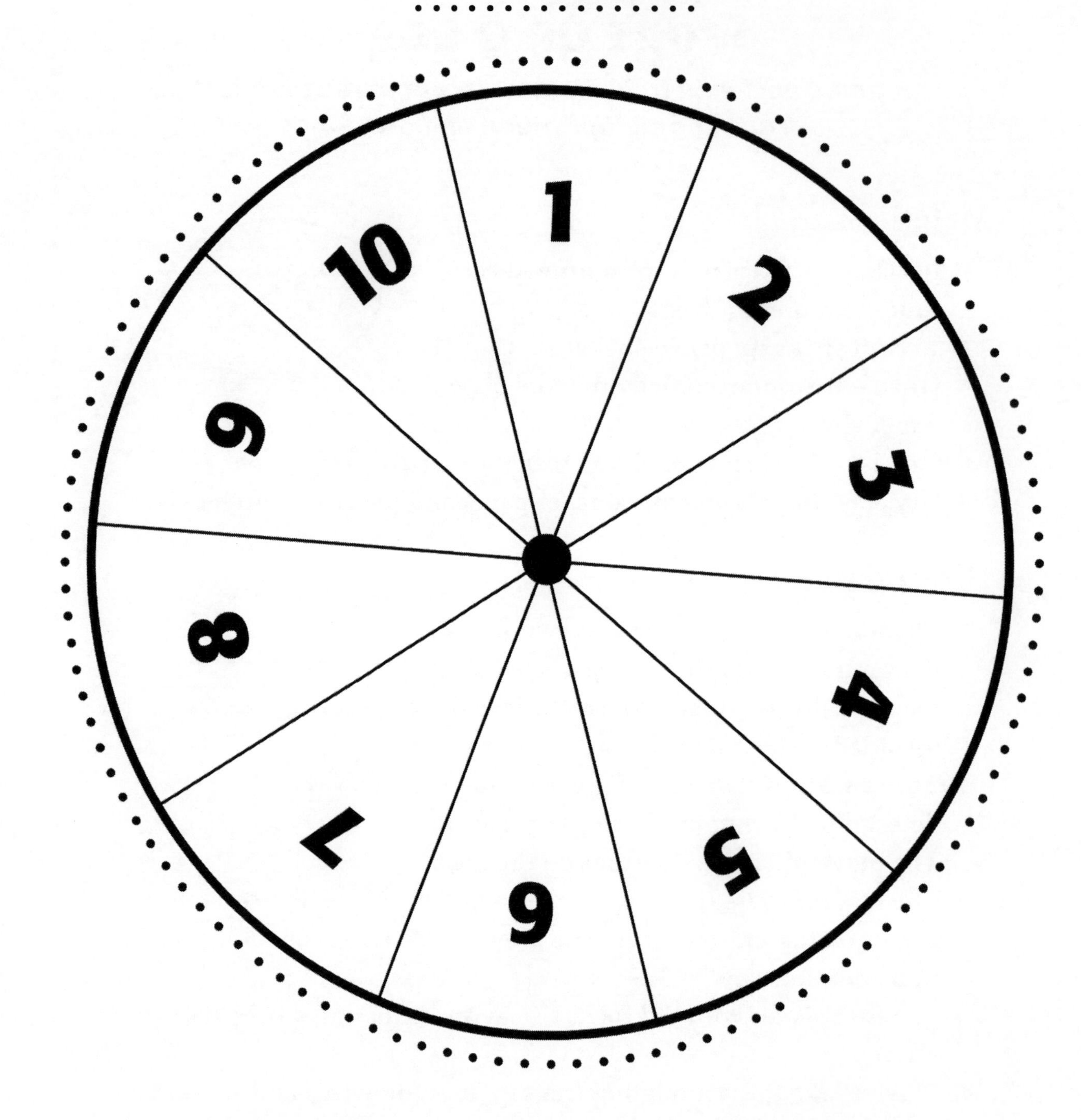

To use the spinner, place a paper clip in the center. Place pencil point through the paper clip at the center of the spinner. Holding the pencil securely with one hand, spin the paper clip with the other hand.

CONCEPT:

Multiplying and Dividing by Ten

What is the Multiplying and Dividing by Ten Concept?

Multiplying and dividing by ten builds fluency with number relationships and computation. When students understand the multiplying and dividing by ten concept, they are more apt to use mental math to solve problems and verify answers. Initially, students multiply ten by single-digit numbers and use 2-digit dividends to divide by ten. Building on this idea, they can learn to work with numbers having greater than 2-digits.

CCSS

Operations and Algebraic Thinking
Number and Operations in Base Ten

Formative Assessment

To find out if a student can multiply and divide by ten ask the student to solve the following equations:

$2 \times 10 =$
$4 \times 10 =$
$5 \times 10 =$

$40 \div 10 =$
$60 \div 10 =$
$90 \div 10 =$

Ask the student to explain how he solved each problem. Analyze the student's level of proficiency through their accuracy, speed, and explanations.

Successful Strategies

One of the foundational concepts associated with multiplying and dividing by ten is rational counting by tens. Encourage students to skip count by tens starting with zero to help them learn the multiples of ten. To further help students understand the multiplying and dividing by ten concept, use base ten blocks or other manipulatives to model the operations.

Math Words to Use

tens, multiply, multiplication, divide, division, times sign, division sign

Questions at Different *Levels of Cognitive Demand*

Recall: *What is 5 x 10?*

Comprehension: *How do you multiply a number by ten?*

Application: *How would you use 7 x 10 to solve 70 ÷ 10?*

Analysis: *Which facts are not directly related to multiples of ten?*

Evaluation: *How would you recommend that someone learn how to multiply and divide by ten?*

Synthesis: *How could you use the multiplying and dividing by ten concept to multiply and divide by nine?*

COMPLEX

Rigorous Problem Solving with the Concept of Multiplying and Dividing by Ten 1

In the game *Turtle Race* players roll numbers 1–9, multiply each number they roll by ten, and move to each product on the game board.

Part A

If Danielle rolls 5, 3, 9, and 7 which products would she move to on the game board?

o 55, 33, 99, and 77
o 50, 30, 90, and 70
o 5, 3, 9, and 7
o 25, 15, 45, and 35
o 15, 13, 19, and 17

Part B

If Danielle's final moves on the game board were 70, 50, 60, which numbers did she roll to win?

☐ ☐ ☐

Justify your answer with division equations.

Rigorous Problem Solving with the Concept of Multiplying and Dividing by Ten 2

In the game *Turtle Race* players roll numbers 1–9, multiply each number they roll by ten, and move to each product on the game board.

Part A

If Kayla rolls 6, 4, 9, and 8 which products would she move to on the game board?

o 6, 4, 9, and 8
o 66, 44, 99, and 88
o 30, 20, 45, and 40
o 60, 40, 90, and 80
o 16, 14, 19, and 18

Part B

If Kayla's final moves on the game board were 80, 20, 30, which numbers did she roll to win?

☐ ☐ ☐

Justify your answer with division equations.

☐

☐

☐

Turtle Race

A game designed to build the concept of multiplying and dividing by ten

Materials:

- Decahedron die (labeled 0–9) or 0–9 spinner
- Turtle Race Game Board (Multiplication)
- 2 small plastic turtles (or cubes to represent turtles)

Directions:

1. Each player places a turtle on one of the start boxes.
2. The first player rolls the die (or spins spinner). The number showing is multiplied by ten. The player can move her turtle if the product is in a box that is touching the box that the turtle is currently occupying. Moves can be forward, backward, vertical, horizontal, or diagonal. If the player cannot move, she misses her turn. If a move is possible, the player must move her turtle (even backwards).
3. The next player rolls the die (or spins the spinner) and moves his turtle, if possible. Both turtles can occupy the same box at the same time, if necessary.
4. Players take turns. The winner is the first turtle to reach the finish line.

Content Differentiation:

Moving Back: Use a hundred chart as a reference. Model equations with base ten blocks.

Moving Ahead: Play the divide by ten version of this game by using numeral cards (0, 10, 20, 30, 40, 50, 60, 70, 80, 90) instead of die or spinner with the Turtle Race (Division) game board. Or add zero to each value on the multiplication game board and multiply by 100.

0–9 spinner

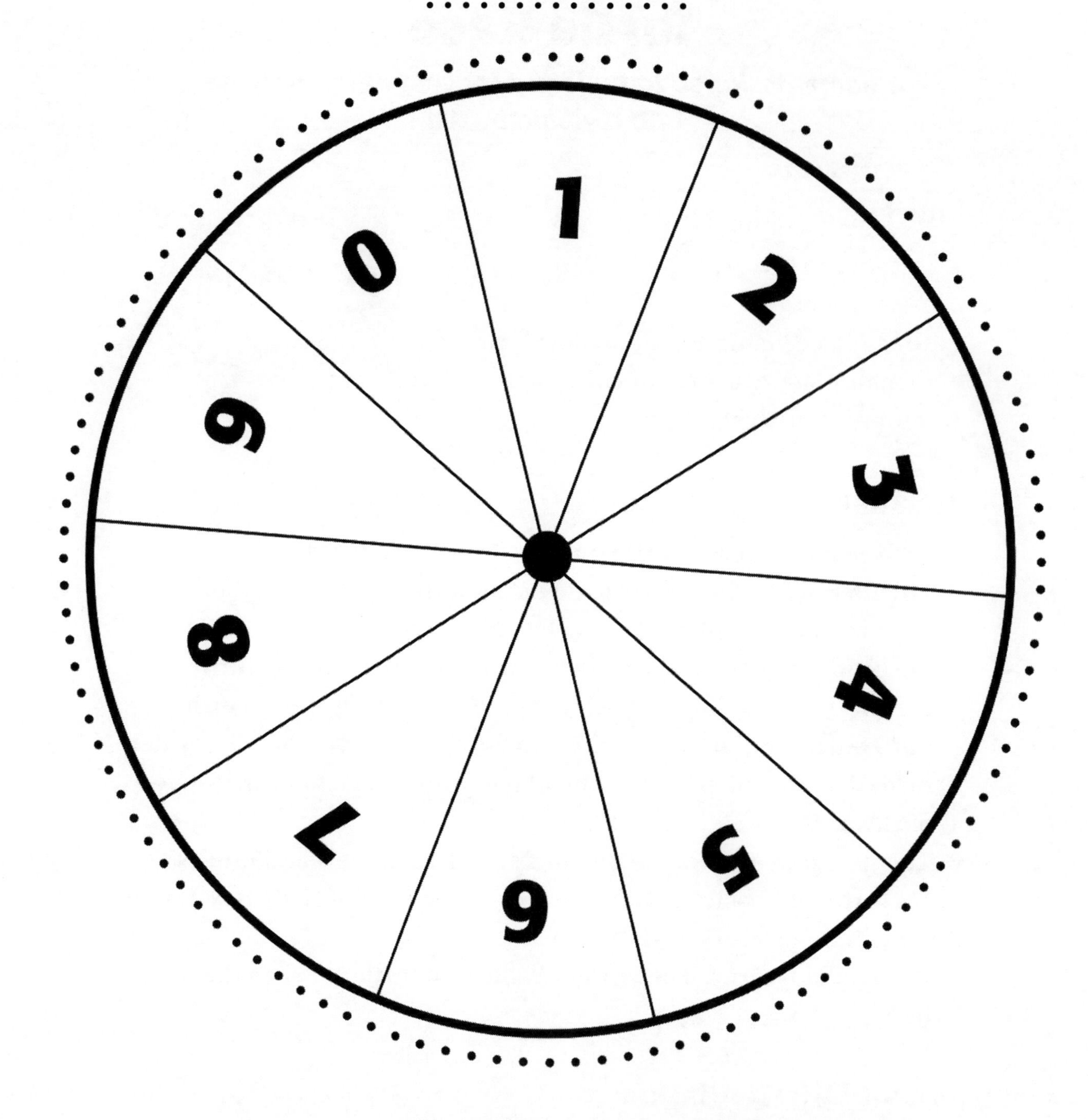

To use the spinner, place a paper clip in the center. Place pencil point through the paper clip at the center of the spinner. Holding the pencil securely with one hand, spin the paper clip with the other hand.

Turtle Race Game Board (Multiplication)

0	Start	30	0	70	40	10	Start	40
10	70	40	10	80	50	20	80	50
20	80	50	20	90	60	30	90	60
30	90	60	30	0	70	40	0	70
40	0	70	40	10	80	50	10	80
50	10	80	50	20	90	60	20	90
60	20	90	60 Finish Line	30 Finish Line	0 Finish Line	70	30	0

Turtle Race Game Board (Division)

go turtle!

0	Start	3	0	7	4	1	Start	4
1	7	4	1	8	5	2	8	5
2	8	5	2	9	6	3	9	6
3	9	6	3	0	7	4	0	7
4	0	7	4	1	8	5	1	8
5	1	8	5	2	9	6	2	9
6	2	9	6 Finish Line	3 Finish Line	0 Finish Line	7	3	0

CONCEPT:

Perfect Squares and Near Squares

What is the Perfect Squares and Near Squares Concept?

As with adding doubles, there is something uniquely fascinating about multiplying a number with itself. Often students' levels of interest are elevated, prompting them to master perfect squares earlier than other multiplication facts. The smaller squares (1 x 1, 2 x 2, 3 x 3, 4 x 4, 5 x 5) and interesting squares (such as 10 x 10) are commonly learned quickly. Sometimes the larger squares (6 x 6, 7 x 7, 8 x 8, 9 x 9, 11 x 11, 12 x 12) take more time. Students require many experiences with perfect squares. When students own these facts, they are ready to use this knowledge to solve near squares multiplication problems. When teaching the near squares concept, encourage students to identify the two perfect squares that the near square falls between and to use plus or minus one group to solve the equation. For example, the near square 8 x 7 = 56 falls between 7 x 7 = 49 and 8 x 8 = 64. If a student uses 7 x 7, he will add one more group of seven (49 + 7 = 56). If he uses 8 x 8, he will subtract one group of eight (64 – 8 = 56). Rather than providing rules for students to follow, allow them to decide which perfect square fact they want to use to solve the near square problem so that they understand the math more completely.

CCSS

Operations and Algebraic Thinking

Formative Assessment

Ask the students to solve several squares (3 x 3, 5 x 5, 7 x 7, 9 x 9) and near squares (2 x 3, 5 x 4, 6 x 7, 7 x 8) multiplication problems. Analyze the level of ownership by the students' accuracy and speed.

Successful Strategies

Teaching students to identify perfect squares and near squares is the first step in ownership of these facts. Ask students to explain which two squares "sandwich" the near square. Use manipulatives and graph paper arrays to model one more group and one less group. Write the facts in order so students can see the patterns. As students gain ownership of perfect squares and near squares, we need to encourage them to use this knowledge with related division facts.

Math Words to Use

perfect squares, near squares, product, multiply, multiplication, groups, sets, multiplier, multiplicand, facts

Questions at Different *Levels of Cognitive Demand*

EASY

Recall: *What is 5 x 5?*

Comprehension: *What example could you give of a perfect square multiplication fact?*

Application: *How would you show near squares products with graph paper arrays?*

Analysis: *Which factor of the near square fact do you plan to use?*

Evaluation: *How would you teach near squares to someone who does not understand it?*

Synthesis: *How could you create a perfect squares and near squares game?*

COMPLEX

Rigorous Problem Solving with the Concept of Perfect Squares and Near Squares 1

Joey and James are playing *X Marks the Spot*. In this game, players roll a number, multiply it by itself, and mark the product on the game board. Joey rolled 5, 4, and 7. James rolled 3, 6, and 8.

Part A

Which set of numbers represent Joey's products?

o 5, 4, and 7
o 10, 8, and 14
o 15, 14, and 17
o 50, 40, and 70
o 8, 10, and 15
o 50,40, and 70
o 25, 16, and 49

Part B

Kolby says 6 x 7 = 37 because 6 x 6 = 36 and 36 + 1 = 37. Do you agree with Kolby?

Why or why not?

Rigorous Problem Solving with the Concept of Perfect Squares and Near Squares 2

Kolby and Kirk are playing *X Marks the Spot*. In this game, players roll a number, multiply it by itself, and mark the product on the game board. Kolby rolled 3, 5, and 9. Kirk rolled 4, 7, and 8.

Part A

Which set of numbers represent Kirk's products?

o 4, 7, and 8
o 14, 17, and 18
o 44, 77, and 88
o 40, 70, and 80
o 8, 14, and 16
o 16, 49, and 64

Part B

Joey says 7 x 8 = 56 because 7 x 7 = 49 and 49 + 7 = 56. Do you agree with Joey?

Why or why not?

X Marks the Spot (Multiplication)

A game designed to build the concept of perfect squares and near squares

Materials:

- Decahedron die (labeled 0–9) or 0–9 spinner *(see page 142)*
- X Marks the Spot Game Board for Perfect Squares Products
- Two highlighters or light markers (different colors)

Directions:

1. Players decide which color highlighter each person will use.
2. The first player rolls the die or spins the spinner. The player multiplies the number with itself and looks for the perfect square product on the game board.
3. The player marks the product with a dot using the highlighter to "hold" the spot.
4. The next player rolls or spins a number, multiplies the number with itself and looks for the perfect square product on the game board. The player can mark the product with a dot to hold it OR the player can "claim" any product that is currently being held with a dot. If the player claims a held product, she uses the highlighter to draw X on the product.
5. Players take turns choosing numbers, multiplying, and holding or claiming products. *Note: All products must be held before they are claimed and can be claimed by any player.*
6. The winner is the first player to have four claimed products (Xs) in a row (horizontal, vertical, or diagonal).

Content Differentiation:

Moving Back: Play the game without holding prior to claiming.

Moving Ahead: Play the same game using Near Squares Products instead of Square Products. The near square is one more group or one less group than a perfect square.

X Marks the Spot Game Board for Perfect Squares Products

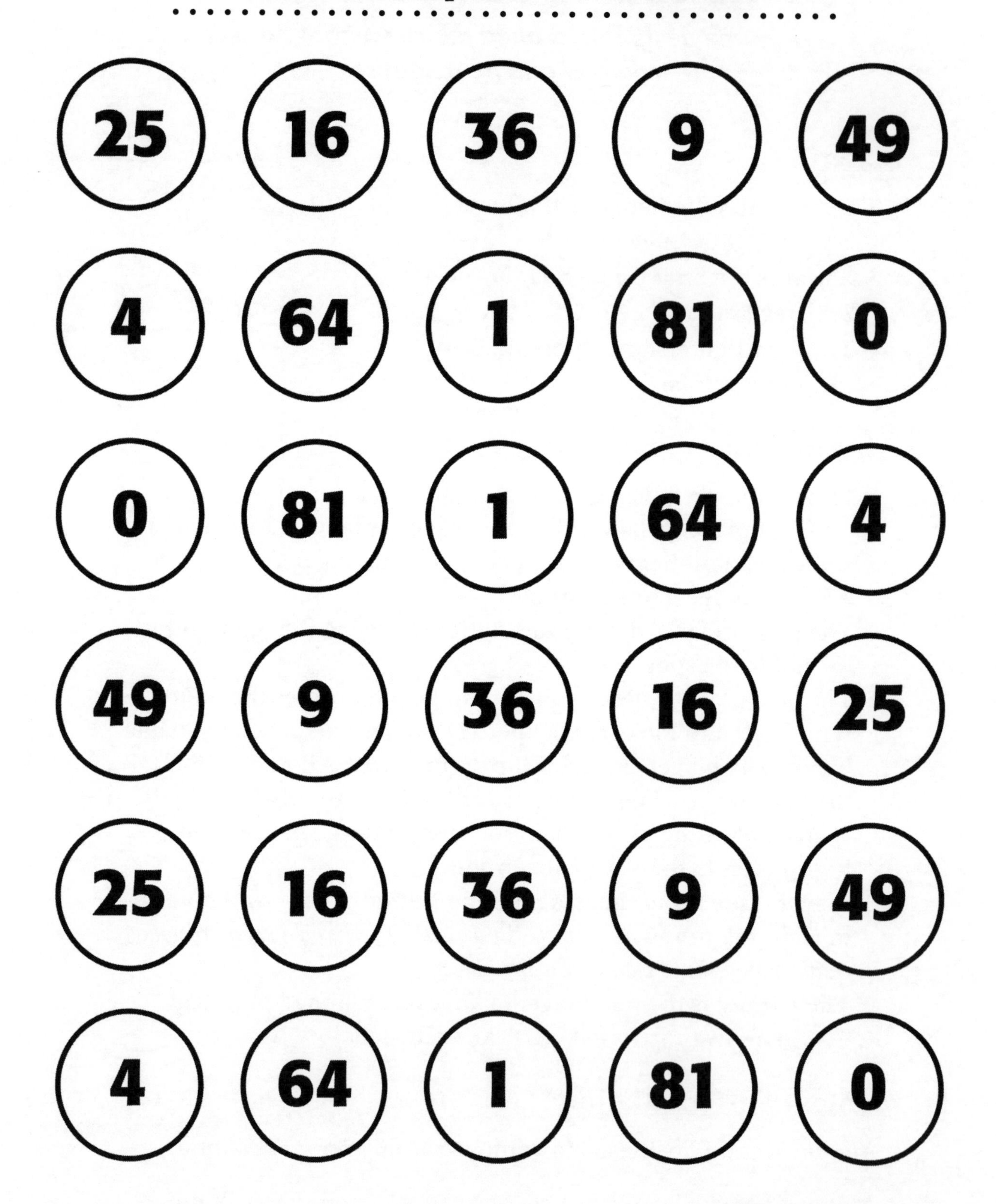

X Marks the Spot Game Board for Near Squares Products

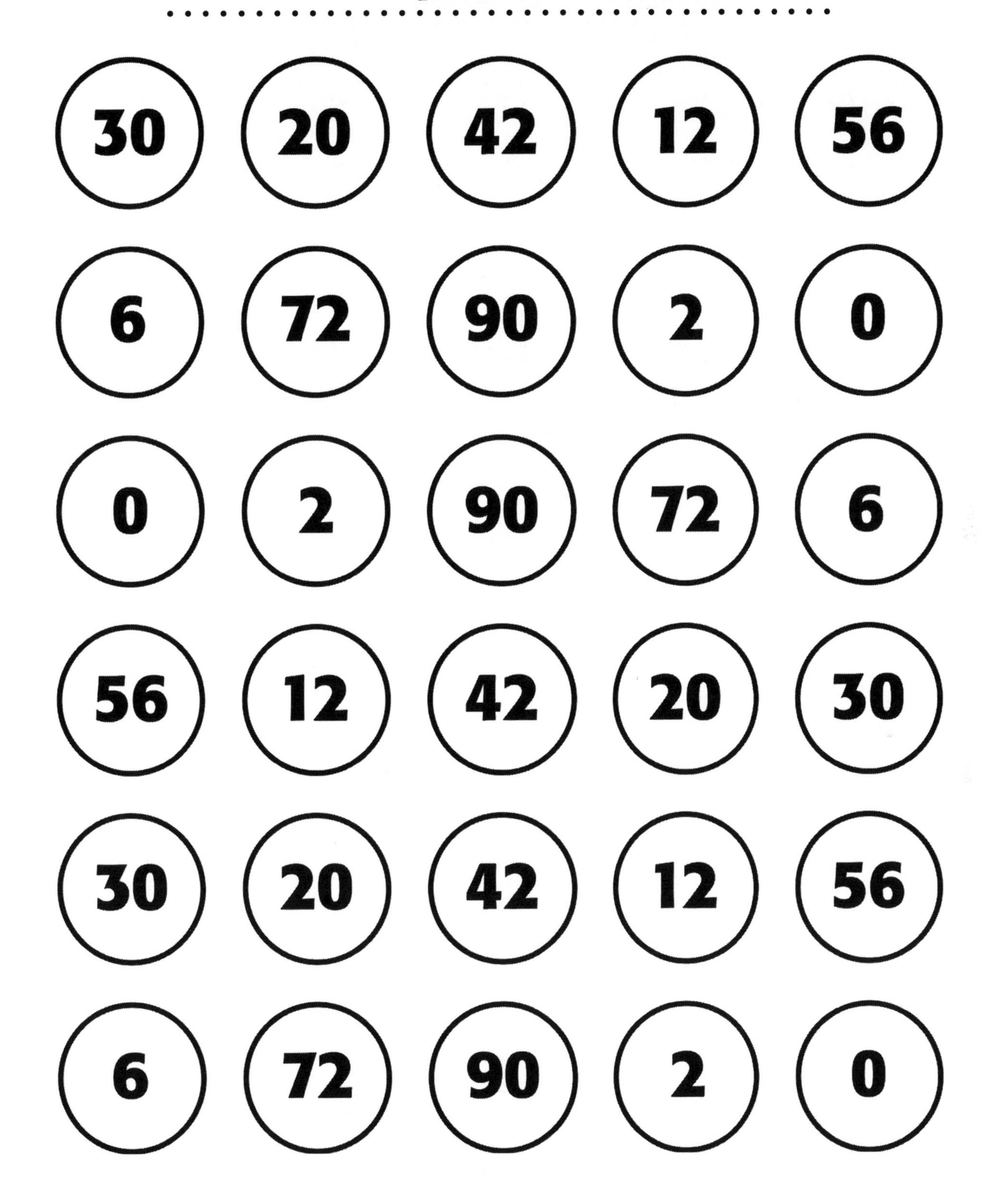

CONCEPT:

Near Tens for Multiplication

What is the Near Tens for Multiplication Concept?

The near tens concept involves adjusting multiplication facts based on their proximity to tens facts. The near tens concept can be used with the nines facts, for example, 9 x 7 = (10 x 7) – (1 x 7). If can also be used with eights facts, for example, 8 x 7 = (10 x 7) – (2 x 7). Additionally, the near tens concept can be used with the elevens facts, for example, 11 x 7 = (10 x 7) + (1 x 7) and the twelves facts, for example, 12 x 7 = (10 x 7) + (2 x 7). While the same type of adjustments can be applied to any facts, the near tens concept is most applicable to the 9s, 8s, 11s, and 12s multiplication facts. When students use the near tens concept, they apply the kind of flexibility with numbers which is needed for mental math and deep conceptual understanding.

CCSS

Operations and Algebraic Thinking

Formative Assessment

To find out if a student understands the near tens for multiplication concept, ask the student to solve the following equations:

6 x 10 =	10 x 8 =	11 x 7 =	9 x 11 =
7 x 9 =	9 x 5 =	5 x 12 =	8 x 12 =
4 x 8 =	8 x 6 =		

Have the student explain how he solves each problem. If the student uses the traditional algorithm (successfully or not successfully), ask the student if he knows another way to solve the equations. Invite the student to try the problems using what they know about multiplying tens.

Successful Strategies

To be successful with the near tens multiplication concept, students must have ownership of multiplying and dividing ten. The near tens concept builds upon what students know about multiplying and dividing tens. For many students, using a multiplication chart helps them to see and apply the near tens concept.

Math Words to Use

near ten, multiply, multiplication, divide, division, adjust, compensate

Questions at Different *Levels of Cognitive Demand*

EASY

Recall: *What is 9 x 10?*

Comprehension: *How could you use a near ten to solve 9 x 6?*

Application: *How would you show the near tens concept with manipulatives?*

Analysis: *Which problems invite the near tens concept?*

Evaluation: *When is it ineffective to use the near tens concept?*

COMPLEX

Synthesis: *How would you teach the near tens concept to someone who did not know it?*

Rigorous Problem Solving with the Concept of Near Tens for Multiplication 1

The Taylor Chocolate Factory produces chocolates in boxes. The number of chocolates in each box is based on the type and size of the chocolates. Kayla bags the boxes and adds a label with the total number of chocolates.

Part A

The first bag contained 9 boxes of 8 chocolates. Which equation uses a place value strategy to represent the total number of chocolates?

o $9 \times 10 = (10 \times 10) - (1 \times 10)$
o $9 \times 8 = (10 \times 8) + (1 \times 8)$
o $9 \times 9 = 81 - 8$
o $9 \times 8 = (10 \times 8) - (1 \times 8)$
o $9 \times 8 = (9 \times 9) - 1$

Part B

What if the second bag of chocolates contained 9 boxes of 7 chocolates?

Justify your answer with an equation that uses a place value strategy.

Part C

How many total boxes in both bags?

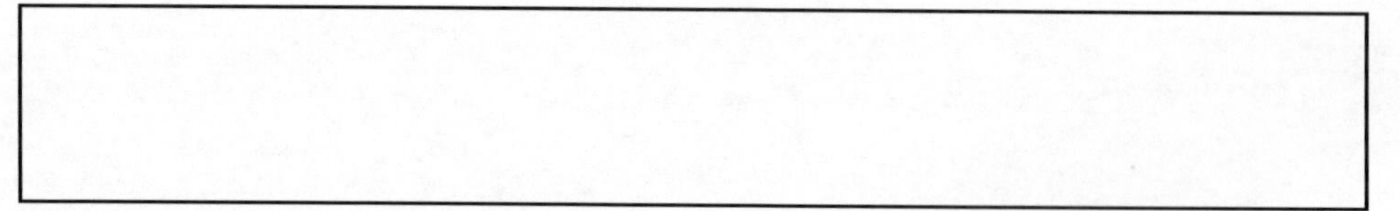

Rigorous Problem Solving with the Concept of Near Tens for Multiplication 2

The Taylor Chocolate Factory produces chocolates in boxes. The number of chocolates in each box is based on the type and size of the chocolates. Kayla bags the boxes and adds a label with the total number of chocolates.

Part A

The first bag contained 9 boxes of 6 chocolates. Which equation uses a place value strategy to represent the total number of chocolates?

- o $9 \times 10 = (10 \times 10) - (6 \times 10)$
- o $9 \times 6 = (10 \times 6) + (1 \times 6)$
- o $9 \times 6 = (10 \times 6) - (1 \times 6)$
- o $9 \times 6 = 90 - 6$
- o $9 \times 6 = (9 \times 9) - 3$

Part B

What if the second bag of chocolates contained 9 boxes of 9 chocolates?

Justify your answer with an equation that uses a place value strategy.

Part C

How many total boxes in both bags?

Roll and Flip

A game designed to build the concept of near tens for multiplication and division

Materials:

- Die with six sides labeled 10 x, 10 x, 9 x, 9 x, 8 x, 8 x
- 0–9 Number Cards *(see page 72)*

Directions:

1. Players shuffle and distribute cards face down.
2. The first player rolls the die and flips one of her cards. If the player rolls 10 x, she multiplies ten times the number on the card. If the player rolls 9 x or 8 x, she uses the 10 times multiplication fact to mentally solve it. For example, 4 x 9 = (4 x 10) – (4 x 1).
3. The player must describe how she uses the near tens concept. The product is the player's score for that round.
4. Players take turns rolling the die, flipping cards, and using near tens to solve the problems.
5. Play continues for ten rounds. Players add their scores from each round. The winner is the player with the highest score.

Content Differentiation:

Moving Back: Use a die with only 10 x and 9 x.

Moving Ahead: Include the related division equations for each round. Change the die to include 11 x and 12 x.

CONCEPT:

Fact Families–Multiplication and Division

What are Fact Families for Multiplication and Division?

Fact families for multiplication and division are facts that are directly related. In the fact family, there are two multiplication problems and two division problems. The multiplication facts are manifestations of the commutative property because changing the order does not change the result of the operation. The division facts are inverse operations of the multiplication facts.

CCSS

Operations and Algebraic Thinking

Formative Assessment

To find out if a student understands multiplication and division fact families, provide the following incomplete fact family and ask the student which fact is missing from the family:

$7 \times 5 = 35$
$35 \div 7 = 5$
$35 \div 5 = 7$

If the student is not successful, try fact families with smaller numbers. If the student is successful, ask him to generate all of the facts in the 6, 9, 54 fact family.

Successful Strategies

Some students struggle with fact families. One of the ways to help these students is to go back to the repeated addition and repeated subtraction concepts. Focusing on the "groups" encourages students to think about the related multiplication and division facts. Additionally, help students to see why squares produce identical facts.

Math Words to Use

multiplication, division, equation, facts, fact families, groups

Questions at Different *Levels of Cognitive Demand*

EASY

Recall: *What is 5 x 4? What is 4 x 5?*

Comprehension: *How would you explain fact families?*

Application: *How would you show the fact family for 6, 8, 48?*

Analysis: *What relationship do you see between 7 x 6 and 42 ÷ 6?*

Evaluation: *What advice would you give someone who does not understand fact families?*

Synthesis: *Is it possible to have a fact family for this equation 2 x 3 x 4 = 24?*

COMPLEX

Rigorous Problem Solving with the Concept of Fact Families—Multiplication and Division 1

In the Fishy Facts game, players try to name the number hidden on one of the fins or the tail and the four facts in the fact family.

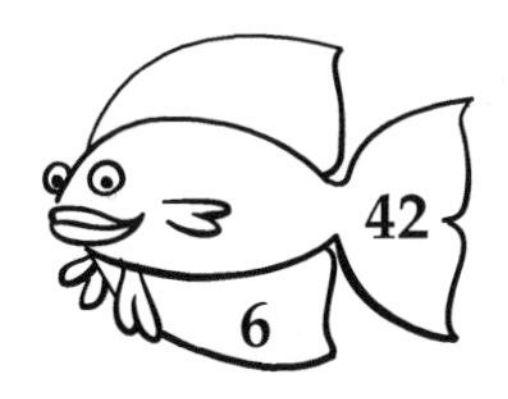

Part A

Which fact belongs to the fact family represented above?

- o $6 \times 6 = 42$
- o $6 \times 8 = 42$
- o $6 \times 7 = 42$
- o $6 \times 5 = 42$
- o $6 \times 9 = 42$

Part B

List the four facts in a different fact family that have a product of 42.

Part C

How would you explain fact families to someone who does not know what they are?

Rigorous Problem Solving with the Concept of Fact Families—Multiplication and Division 2

In the Fishy Facts game, players try to name the number hidden on one of the fins or the tail and the four facts in the fact family.

Part A

Which fact belongs to the fact family represented above?

o $6 \times 6 = 48$
o $6 \times 8 = 48$
o $6 \times 7 = 48$
o $6 \times 5 = 48$
o $6 \times 9 = 48$

Part B

List the four facts in a different fact family that have a product of 48.

Part C

Why are fact families important?

Fishy Facts

A game designed to build the concept of multiplication zand division fact families

Materials:

- Fishy Facts resource pages, scissors
- Bowl

Directions:

1. Cut out fish and place in a bowl.
2. One player reaches into the bowl and pulls out one fish. The player displays the fish to other players.
3. When displaying a fish, the player holds the fish by the tail or one of the fins to cover up one of the numbers in the fact family.
4. The first player to correctly identify the covered number has an opportunity to earn the fish by naming all four facts in the fact family.
5. Players take turns displaying fish, identifying covered numbers, and naming facts in fact families.
6. The winner is the player with the most fish after all (or a specified number) of the fish are displayed.

Content Differentiation:

Moving Back: Use selected fish (2s, 5s, 10s).

Moving Ahead: Use the blank fish to make different fact families.

Fishy Facts

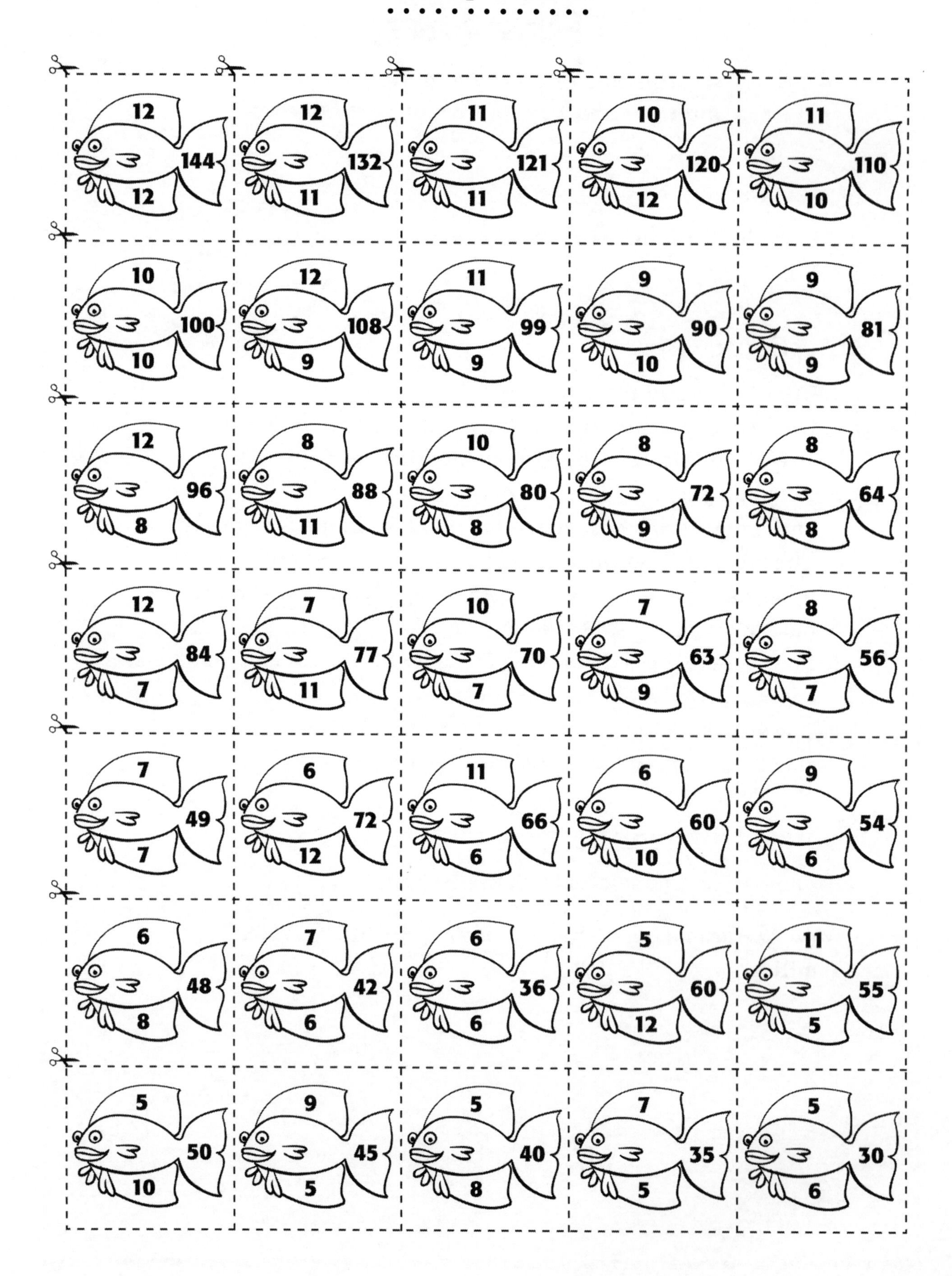

12
144
12
12
132
11
11
121
11
10
120
12
11
110
10
10
100
10
12
108
9
11
99
9
9
90
10
9
81
9
12
96
8
8
88
11
10
80
8
8
72
9
8
64
8
12
84
7
7
77
11
10
70
7
7
63
9
8
56
7
7
49
7
6
72
12
11
66
6
6
60
10
9
54
6
6
48
8
7
42
6
6
36
6
5
60
12
11
55
5
5
50
10
9
45
5
5
40
8
7
35
5
5
30
6

Fishy Facts

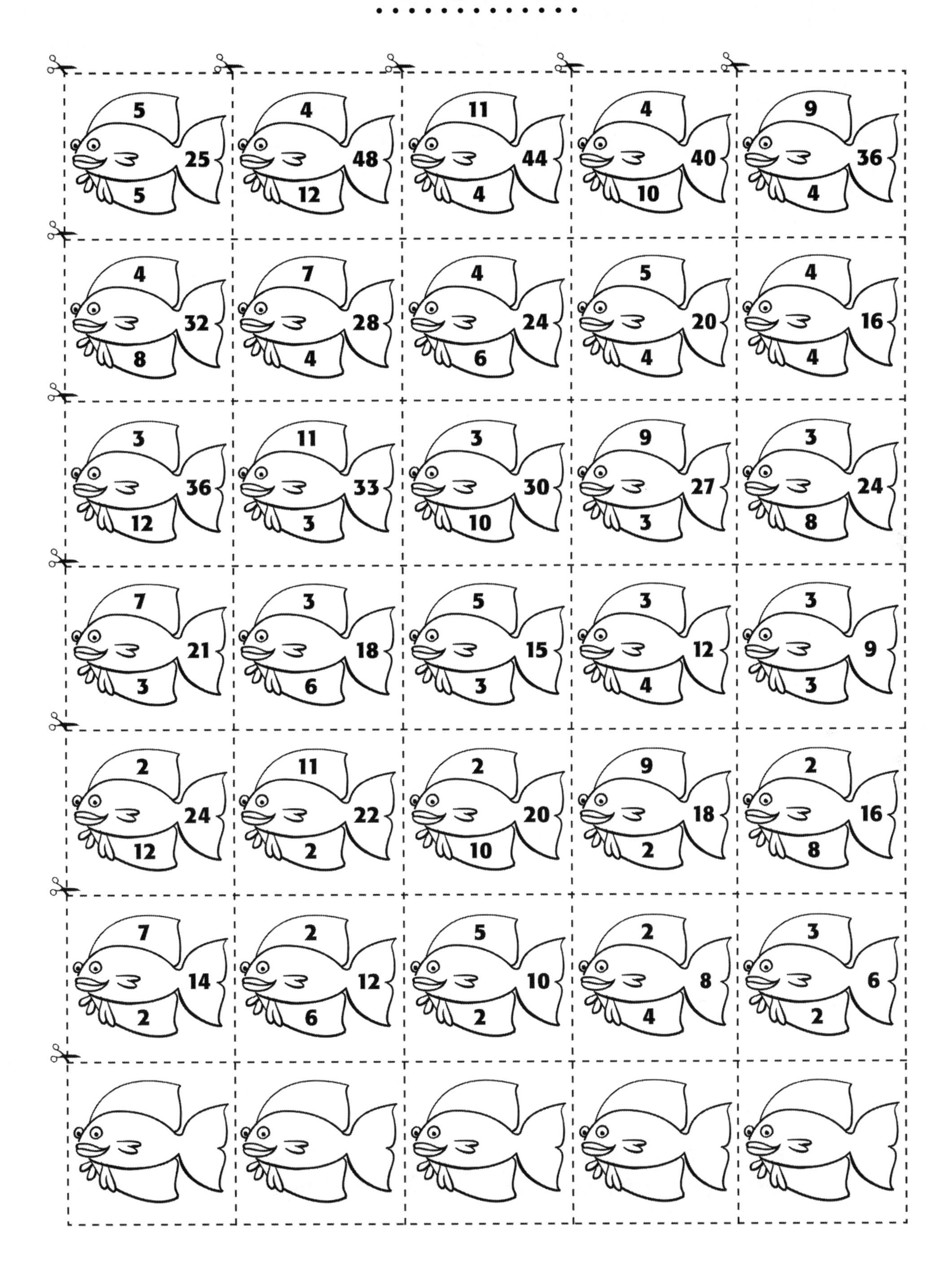

5
25
5
4
48
12
11
44
4
4
40
10
9
36
4
4
32
8
7
28
4
4
24
6
5
20
4
4
16
4
3
36
12
11
33
3
3
30
10
9
27
3
3
24
8
7
21
3
3
18
6
5
15
3
3
12
4
3
9
3
2
24
12
11
22
2
2
20
10
9
18
2
2
16
8
7
14
2
2
12
6
5
10
2
2
8
4
3
6
2

CONCEPT:

Partial Products

What is the Partial Products Concept?

The partial products concept is a multiplication strategy that involves breaking down a larger problem into smaller, more manageable problems. The traditional algorithm for multiplication involves procedures, such as aligning digits in columns and always starting with the ones. For many students the rules of the traditional algorithm are meaningless. They cannot remember what to do because they do not understand the math. Alternatives to algorithms such as partial products help build students' conceptual understanding. With partial products, it does not matter if the columns are aligned or if the student starts with the ones. There is no need to "bring down a zero" or "put the big number on top." Let's look at 6 x 14. Using the distributive property, we break it down into smaller problems, 6 (10 + 4) = (6 x 10) + (6 x 4). Here, 60 is part of the product and 24 is part of the product. These partial products are easily combined into the total product of 84.

CCSS

Operations and Algebraic Thinking
Number and Operations in Base Ten

Formative Assessment

To find out if a student understands the partial products concept ask the student to solve the following equations:

13 x 4 = 26 x 5 = 52 x 7 =

16 x 12 = 18 x 27 = 46 x 27 =

Have the student explain how he solves each problem. Invite the student to try the problems without aligning the digits in columns and without starting with the ones.

Successful Strategies

One of the reasons students struggle with the traditional algorithm is the inaccurate descriptions of the procedures they have heard. Unfortunately, many people use digit names rather than number values, creating potential misconceptions. When solving 21 x 6, they may say 6 x 1 and 6 x 2. The accurate description is 6 x 1 and 6 x 20. The partial products concept is embedded in place value and conceptual understanding, which help remedy algorithmic misconceptions. Even if students successfully use the traditional algorithm, learning the partial products concept strengthens conceptual understanding, increases mental math capacity, and improves flexibility and fluency.

Math Words to Use

partial, part, product, multiplication, multiply, equation, equal, distributive property

Questions at Different *Levels of Cognitive Demand*

Recall: *What is 27 x 8?*

Comprehension: *How did you solve the problem?*

Application: *What picture could you draw to show the partial products?*

Analysis: *How would you compare the partial products?*

Evaluation: *Which products are easier to work with? Why?*

Synthesis: *How would you design a template for recording partial products?*

COMPLEX

Rigorous Problem Solving with the Concept of Partial Products 1

Lanham Elementary School received a donation of juice boxes from the local store. Each class was given 43 juice boxes. There were 27 classes. How many total juice boxes did the school receive?

Part A

Which expressions represent the total number of juice boxes the school received?

- o 43 + 27
- o 21 + 280 + 60 + 800
- o 21 + 28 +60 + 80
- o (20 x 40) + (20 x 3) + (7 x 40) + (7 x 3)
- o (2 x 40) + (20 x 3) + (7 x 4) + (7 x 3)
- o (2 x 40) + (20 x 3) + (7 x 40) + (7 x 3)

Part B

What if each class received 4 more juice boxes?

Write an expression to represent the new total number of juice boxes the school received.

Rigorous Problem Solving with the Concept of Partial Products 2

Aloysius Elementary School received another donation of juice boxes from the local store. Each class was given 57 juice boxes. There were 32 classes. How many total juice boxes did the school receive?

Part A

Which expressions represent the total number of juice boxes the school received?

- o 57 + 32
- o (30 x 50) + (30 x 7) + (2 x 50) + (2 x 7)
- o 14 + 100 + 210 + 1500
- o (3 x 50) + (30 x 7) + (2 x 50) + (2 x 7)
- o (30 x 50) + (3 x 7) + (2 x 50) + (2 x 7)
- o 14 + 10 + 21 + 15
- o (30 x 50) + (3 x 7) + (2 x 5) + (2 x 7)

Part B

What if each class received 2 more juice boxes?

Write an expression to represent the new total number of juice boxes the school received.

Products in Parts

A game designed to build the concept of partial products

PLAYERS: 3 OR MORE

Materials:

- Several sheets of centimeter graph paper
- Highlighter for each player
- Base ten blocks
- Two red dice (labeled 1–6 with numerals)
- Two white dice (labeled 1–6 with numerals)
- Dry erase boards and markers (or paper and pencils)

Directions:

1. The first player rolls all four dice to generate a 2-digit times 2-digit multiplication equation. The red dice represent the tens and the white dice represent the ones.
2. The player shows the equation with lines and dots using the highlighter on graph paper. For example, 12 x 21 includes one line of ten and two dots of one *times* two lines of ten and one dot of one.
3. The player shows a model of the partial products by placing base ten blocks to show the array.
4. Other players race to record the partial products (10 x 20) + (10 x 1) + (2 x 20) + (2 x 1) = 200 + 10 + 40 + 2. The first player with the correct partial products scores 1,000 points.
5. Players take turns rolling, modeling and recording partial products. The winner is the first player to score 10,000.

21

12

Content Differentiation:

Moving Back: Try single-digit times 2-digit equations.

Moving Ahead: Use decahedron dice or 0–9 spinners.

CONCEPT: Partial Quotients

What is the Partial Quotients Concept?

The partial quotients concept is a division strategy that involves consecutive approximations. Like partial products, this concept is as an alternative to the traditional algorithm. For some students the rules of long division are difficult to remember and they miss steps resulting in incorrect answers. The beauty of the partial quotients concept is the focus on building conceptual knowledge of division through estimation. Students estimate how many multiples of the divisor they should work with each time they divide. These partial quotients are combined to name the complete quotient. Remainders are easily identified.

$14\overline{)1092}$	
– 700	50
392	
– 280	20
112	
– 70	5
42	
– 42	3
0	
R 0	78 Quotient

CCSS

Operations and Algebraic Thinking

Formative Assessment

To find out if a student understands partial quotients ask the student to solve the following equations:

$5\overline{)125}$ $7\overline{)3360}$ $16\overline{)304}$ $42\overline{)7336}$

Ask the student to explain how she solves each problem. If she uses the traditional algorithm, ask if she knows another way. Encourage the student to use estimation to find parts of the quotient.

Successful Strategies

Encourage students to come up with ways to record the partial quotients. Students may decide to record the partial quotients with a different color marker, on a separate sheet of paper, or on sticky notes. If students figure out what works best for them, they are more likely to know what to do because they understand why it works. As with partial products, even if students successfully use the traditional algorithm, learning the partial quotients concept strengthens conceptual understanding, increases mental math capacity, and improves flexibility and fluency.

Math Words to Use

partial, part, quotient, divide, division, strategy, equation, equal, dividend, divisor

Questions at Different *Levels of Cognitive Demand*

EASY

Recall: *What is 120 divided by five?*

Comprehension: *How would you divide 270 by 45?*

Application: *How do you show a remainder?*

Analysis: *How would you categorize your partial quotients?*

Evaluation: *Which initial partial quotient do you recommend? Why?*

Synthesis: *What if the remainder was divided into fractions?*

COMPLEX

Rigorous Problem Solving with the Concept of Partial Quotients 1

David has a jar of 912 quarters. The jar is full so he decided to separate the quarters into bags. Each bag holds exactly 48 quarters.

Part A

How many bags will he need to hold all of the quarters?

o 8 bags
o 18 bags
o 19 bags
o 80 bags
o 180 bags
o 90 bags
o 190 bags

Part B

How much money is in each bag?

Justify your answers with an equation.

Rigorous Problem Solving with the Concept of Partial Quotients 2

Donald has a jar of 714 quarters. The jar is full so he decided to separate the quarters into bags. Each bag holds exactly 42 quarters.

Part A

How many bags will he need to hold all of the quarters?

- o 7 bags
- o 17 bags
- o 18 bags
- o 80 bags
- o 180 bags
- o 70 bags
- o 170 bags

Part B

How much money is in each bag?

Justify your answers with an equation.

Prove It

A game designed to build the concept of partial quotients

Materials:

- Three blue dice (labeled 1–6)
- Two yellow dice (labeled 1–6)
- Dry erase boards and marker (or paper and pencil) for each team
- 2-minute timer

Directions:

1. The first team rolls all three blue dice. The team arranges the dice in any way to show the 3-digit number that will serve as the dividend.
2. The other team rolls both yellow dice. The team arranges the dice in any way to show the 2-digit number that will serve as the divisor.
3. The timer is set. All teams write the division problem on dry erase boards and work together to solve the equation using partial quotients.
4. Any team that correctly solves the equation using partial quotients in under two minutes earns points for the round. The quotient tells how many points are earned for the round.
5. Before erasing boards for the next round, players must prove their use of partial quotients, compare answers, and discuss which partial quotients were used and why.
6. Teams keep a running total of points earned. After ten rounds, the team with the most points wins.

Content Differentiation:

Moving Back: Use two blue dice and one yellow die.

Moving Ahead: Use four blue dice and three yellow dice.

CONCEPT:

Equal Products

What is the Equal Products Concept?

The equal products concept is a multiplication strategy based on the idea that different multiplication equations can have the same product. This concept involves manipulating the multiplicand and the multiplier in a way that results in the same product. We can double the multiplicand and halve the multiplier (or vice versa) to create a new expression that has the same product, for example, 25 x 6 = 50 x 3. In this and other situations, doubling and halving to produce equal products results in a problem that is easier to mentally solve. The key is to know when this concept is most applicable. Using the equal products concept helps students understand number relationships and builds conceptual knowledge of multiplication and division.

CCSS

Operations and Algebraic Thinking
Number and Operations in Base Ten

Formative Assessment

To find out if a student understands the equal products concept, ask the student to solve the following equations:

15 x 2 =   25 x 12 =   35 x 6 =   75 x 16 =

Ask the student explain how he solves each problem. If the student uses the traditional algorithm (successfully or not successfully), ask him if he knows another way to solve the equations. Encourage the students to solve the problems using what they know about doubling and halving.

Successful Strategies

While the equal products concept can be used in any multiplication situation, it is most efficient when the newly created expression invites mental math. We need to teach students to get a feel for the numbers they are working with. Is the multiplicand (or multiplier) a number that is easier to work with when it is doubled? Is the multiplier (or multiplicand) even? Students must learn how to analyze multiplication equations to help them figure out which concept is preferred. For some students the equal products concept is a mystery. If they do not understand why the doubling and halving procedures work, they will not build conceptual understanding of multiplication. One of the best ways to model the equal products concept is with arrays. If we have a paper array that is 25 x 6, we can cut the array and slide part of it to show 50 x 3.

Math Words to Use

product, multiplicand, multiplier, multiplication, multiply, equal, equation, expression

Questions at Different *Levels of Cognitive Demand*

Recall: *What is 25 x 8?*

Comprehension: *How would you solve 35 x 6?*

Application: *Where would you cut and how would you move the array to show an equal product?*

Analysis: *How would you compare and contrast the products?*

Evaluation: *Is equal products always the most efficient way to solve an equation? Why?*

COMPLEX

Synthesis: *What if we quadrupled the multiplier?*

Rigorous Problem Solving with the Concept of Equal Products 1

Kayla and her Grandmother made holiday cookies. Each tray had 35 cookies on it. When Grandmother took the sixth tray out of the oven, Kayla asked, "How many cookies did we bake?"

Part A

Grandmother said, "We have 6 groups of 35, but let's half the groups and double the cookies in each group by putting the trays in sets of 2."

Which equation represents the equal products?

- o 6 x 35 = 2 x 70
- o 6 x 35 = 12 x 70
- o 6 x 35 = 12 x 35
- o 6 x 35 = 3 x 70
- o 6 x 35 = 12 x 7

Part B

Joey said, "If there were 12 trays of 75 cookies we could half the trays and double the cookies twice which would be 3 x 300." Do you agree or disagree with Joey?

Justify your answers with equations.

Rigorous Problem Solving with the Concept of Equal Products 1

Sophia and her Grandmother made holiday cookies. Each tray had 55 cookies on it. When Grandmother took the fourth tray out of the oven, Sophia asked, "How many cookies did we bake?"

Part A

Grandmother said, "We have 4 groups of 55, but let's half the groups and double the cookies in each group by putting the trays in sets of 2."

Which equation represents the equal products?

o $4 \times 55 = 2 \times 100$
o $4 \times 55 = 2 \times 55$
o $4 \times 55 = 4 \times 110$
o $4 \times 55 = 8 \times 11$
o $4 \times 55 = 2 \times 110$

Part B

Kirk said, "If there were 8 trays of 45 cookies we could half the trays and double the cookies twice which would be 2 x 180." Do you agree or disagree with Kirk?

Justify your answers with equations.

That's It

A game designed to build the concept of equal products

Materials:

- 15–95 Spinner copied on yellow paper, paper clip
- 2–18 Spinner copied on pink paper, paper clip
- Timer and calculator

Directions:

1. The first player spins the yellow spinner to find the multiplicand and spins the pink spinner to find the multiplier.
2. The player states the multiplication equation while the other player starts the timer. The player doubles the multiplicand, halves the multiplier, and states the equal products. For example the player says, "45 x 4 = 90 x 2. That's it! 180."
3. The other player stops the timer. The number of seconds is recorded as the player's score. Players keep a running total throughout the game.
4. After each player has one turn the round is complete. After each round, the spinners alternate. The pink spinner is the multiplicand and the yellow spinner is the multiplier.
5. Players take turns spinning, creating, and stating equal products, timing each other, and verifying answers with the calculator.
6. The winner is the player with the lowest score after ten rounds.

Content Differentiation:

Moving Back: Using a graph paper array for each equation, cut and move the array to model the doubling and halving.

Moving Ahead: Make a new pink spinner with larger even numbers.

15–95 Spinner

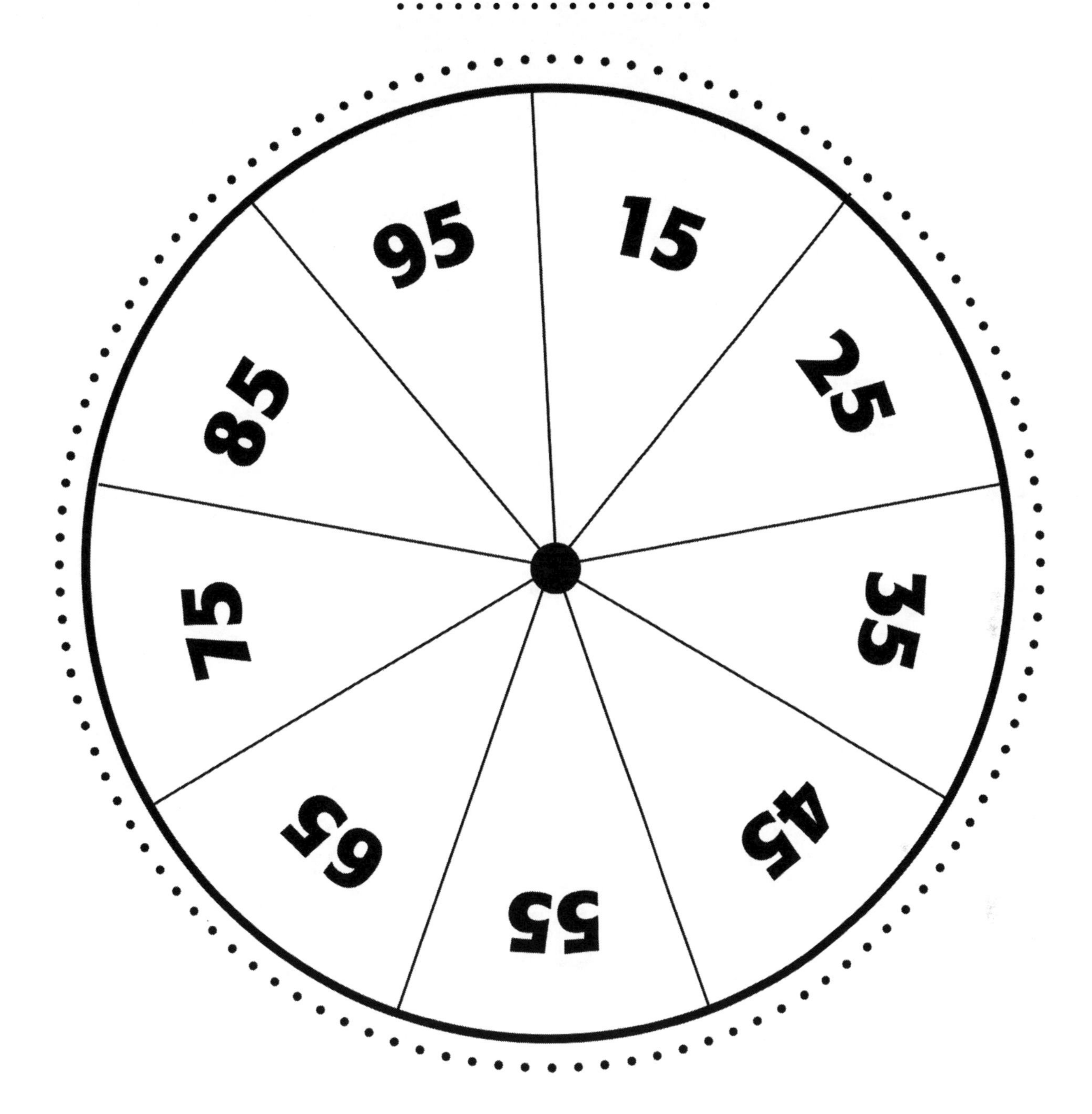

To use the spinner, place a paper clip in the center. Place pencil point through the paper clip at the center of the spinner. Holding the pencil securely with one hand, spin the paper clip with the other hand.

2–18 Spinner

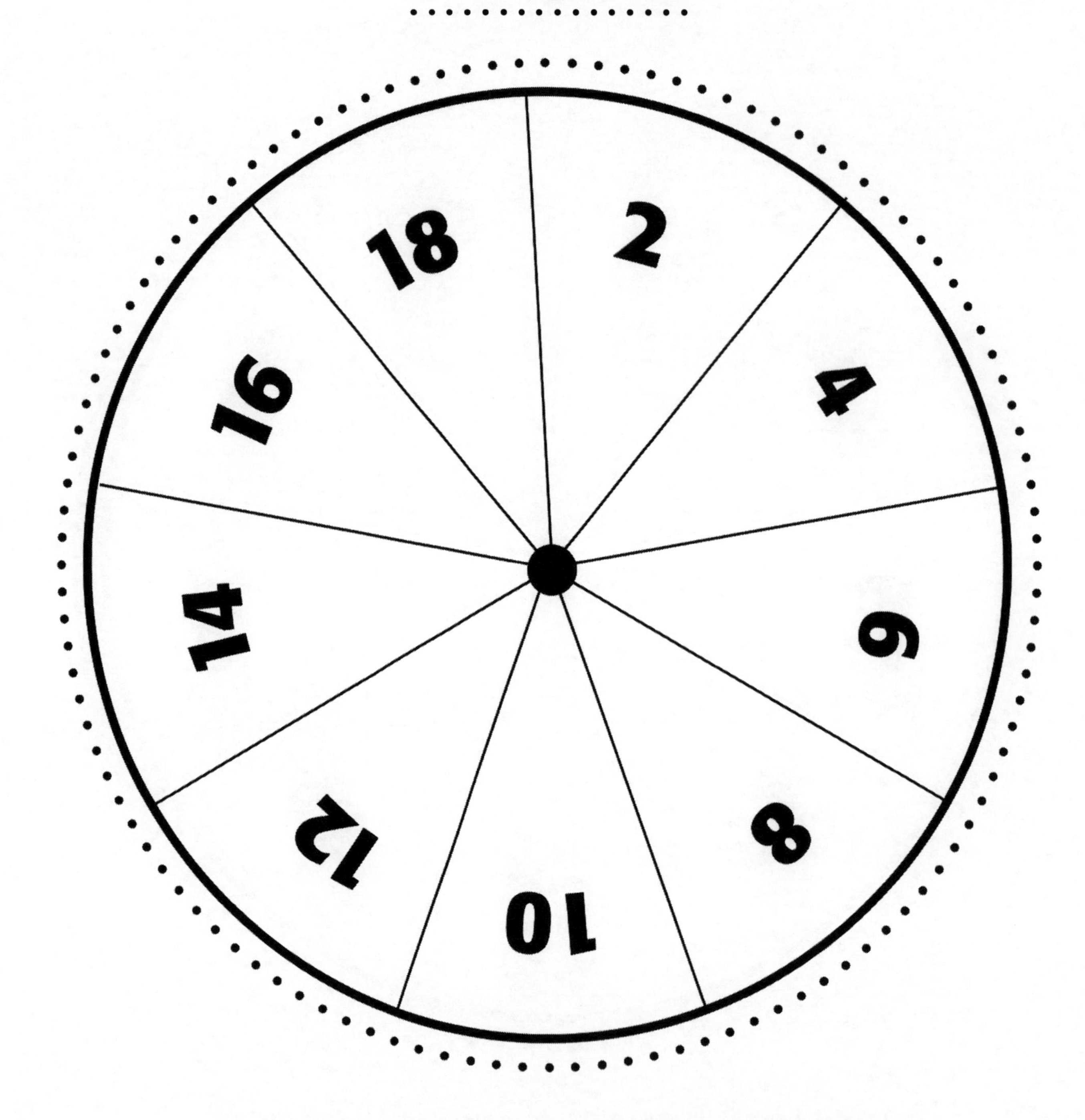

To use the spinner, place a paper clip in the center. Place pencil point through the paper clip at the center of the spinner. Holding the pencil securely with one hand, spin the paper clip with the other hand.

CHAPTER

Multifaceted Number Concepts

Expanded Form

Thousands and Millions

Commutative and Associative Properties

Prime and Composite Numbers

Finding Common Multiples

Finding Common Factors

CONCEPT:
Expanded Form

What is the Expanded Form Concept?

Expanded form involves expressing a number as a sum of the products of each digit. The expanded form concept is deeply rooted in place value. Expressing the value of any number by highlighting the parts of that number helps students build fluency with numbers. Working with expanded form also helps students understand how numbers are composed and decomposed.

CCSS

Number and Operations in Base Ten

Formative Assessment

To find out if a student understands expanded form, ask the student to name the values for the following expanded form expressions:

(4 x 10) + (8 x 1)	(3 x 100) + (5 x 10) + (3 x 1)
(7 x 100) + (6 x 1)	(2 x 1,000) + (3 x 100) + (9 x 10) + (1 x 1)
(7 x 1,000) + (8 x 100) + (3 x 1)	(4 x 10,000) + (6 x 100) + (5 x 10)

Ask the student to give the expanded form of the following numbers:

63	425
907	4,820
6,039	24,309

Find out the range of the student's knowledge of the expanded form concept by identifying to which place value the student is successful.

Successful Strategies

To help strengthen understanding of the expanded form concept students need to build numbers with place value manipulatives (such as base ten blocks and Digi-Blocks™). As students build numbers, they need to discuss what the values look like and the differences among values within the number. The math discourse combined with the visualization and manipulation is powerful for students.

Math Words to Use

expanded form, digits, values, expression, equation

Questions at Different *Levels of Cognitive Demand*

Recall: *What is (5 x 100) + (3 x 10) + (7 x 1)?*

Comprehension: *What is expanded form?*

Application: *How would you show 3,268 in expanded form?*

Analysis: *How would you compare the parts of expanded form?*

Evaluation: *Why is expanded form a helpful way to show numbers?*

Synthesis: *How would you use expanded form with decimal fractions?*

COMPLEX

Rigorous Problem Solving with the Concept of Expanded Form 1

Kolby rolls four dice and forms a 4-digit number.

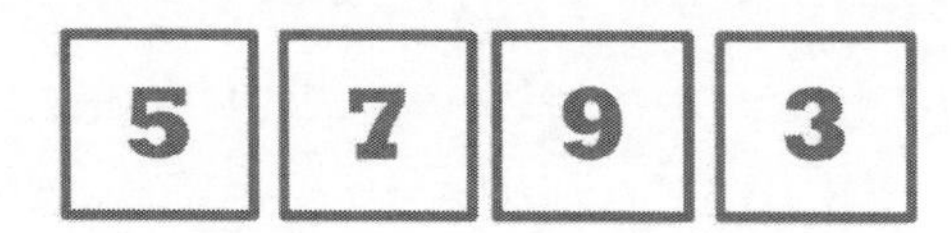

Part A

Which expressions do *not* represent the number in expanded form?

- o $(5 \times 1,000) + (7 \times 100) + (9 \times 10) + (1 \times 3)$
- o $(5 \times 100) + (7 \times 100) + (9 \times 10) + (1 \times 3)$
- o $5,000 + 700 + 90 + 3$
- o $5,000 + 700 + 30 + 9$
- o $(1,000 \times 5) + (100 \times 7) + (10 \times 9) + (3 \times 1)$

Part B

What is the greatest 4-digit number Kolby could have made with the four numbers he rolled?

Write the number and the expanded form as an equation.

What is the smallest 4-digit number Kolby could have made with the four numbers he rolled?

Write the number and the expanded form as an equation.

Rigorous Problem Solving with the Concept of Expanded Form 2

Kirkland rolls four dice and forms a 4-digit number.

7	9	6	8

Part A

Which expressions do *not* represent the number in expanded form?

- o 7,000 + 900 + 60 + 8
- o 7,000 + 900 + 80 + 6
- o (1,000 x 7) + (100 x 9) + (10 x 6) + (8 x 1)
- o (7 x 1,000) + (9 x 100) + (6 x 10) + (1 x 8)
- o (7 x 100) + (9 x 100) + (6 x 10) + (1 x 8)

Part B

What is the greatest 4-digit number Kirkland could have made with the four numbers he rolled?

Write the number and the expanded form as an equation.

What is the smallest 4-digit number Kirkland could have made with the four numbers he rolled?

Write the number and the expanded form as an equation.

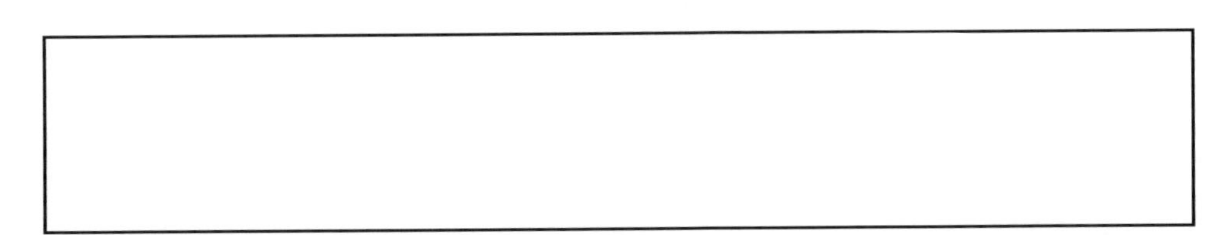

Race Track

A game designed to build the concept of expanded form

Materials:

- 4 dice (labeled 1–6 with numerals) for each player
- Race Track for each player
- Pencil or pen for each player

Directions:

1. Players say, "On your mark. Get set. Go!"
2. Players roll all four dice and form a 4-digit number with the numbers rolled.
3. Players write this number on the first speed limit sign and write the expanded form of this number in the first section of the race track.
4. As soon as the section of the race track is complete, the player rolls again and repeats the process.
5. All players are filling in individual race tracks at the same time.
6. The winner is the first player to accurately complete his race track.
7. Answers are verified by other players after the game.

Content Differentiation:

Moving Back: Use three dice for 3-digit numbers.

Moving Ahead: Use more than four dice for larger numbers.

Race Track

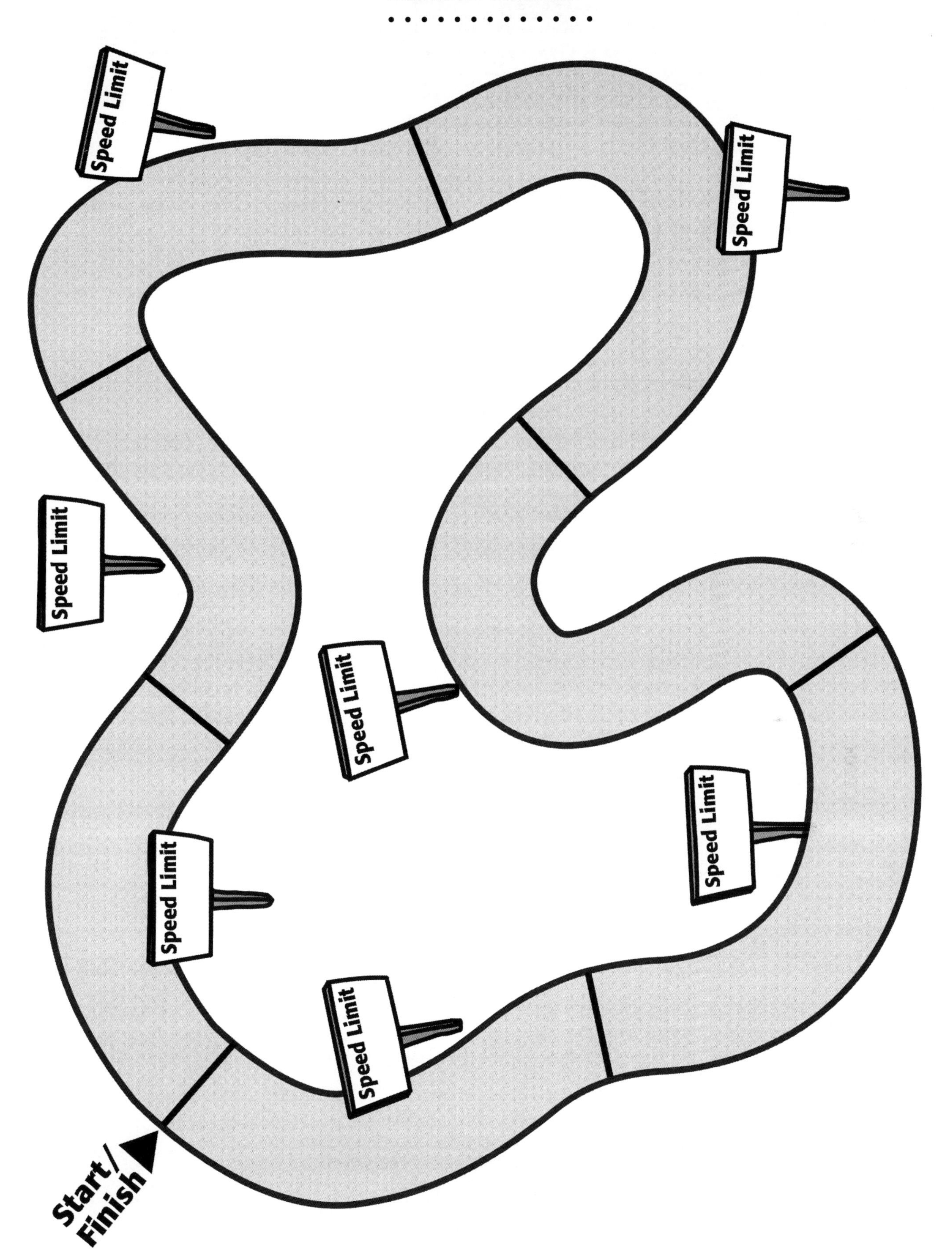
Speed Limit
Speed Limit
Speed Limit
Speed Limit
Speed Limit
Speed Limit
Speed Limit
Start/
Finish

CONCEPT:

Thousands and Millions

What is the Thousands and Millions Concept?

Using a positional system, digits occupy places to name number values. Some students struggle with multi-digit numbers. Students end up reading digits rather than reading the actual number because these large numbers do not have meaning to the students.

Millions			Thousands			Ones		
hundreds	tens	ones	hundreds	tens	ones	hundreds	tens	ones
1	3	5	7	9	2	4	6	8

The number is *one hundred, thirty-five million, seven hundred ninety-two thousand, four hundred sixty-eight*. Provide students with meaningful experiences which include seeing, hearing, and building numbers.

CCSS

Number and Operations in Base Ten

Formative Assessment

To find out if a student understands thousands and millions ask the student to write the following numbers with digits:

One thousand, eight hundred, forty-nine
Two hundred thousand, seven hundred, eight
Five million, six thousand, two hundred, fifty-three

Ask the student to read the following numbers:

4,357 93,618 6,278,021 23,062,340

Successful Strategies

To help students gain comfort with large numbers we can provide them with representations. Ask students to examine 10,000 dots *(see page 192)* to build meaning for the value. By making 100 copies of the 10,000 dots, we can provide a representation of 1,000,000 for students. Additionally, model math language as numbers are read with students.

Math Words to Use

ones, tens, hundreds, thousands, ten thousand, hundred thousand, million, ten million, hundred million

Questions at Different *Levels of Cognitive Demand*

Recall: *Which place is between thousands and tens?*

Comprehension: *How many zeros are in ten thousand?*

Application: *How would you show 20,000?*

Analysis: *How would you compare 10,000 to 100,000?*

Evaluation: *What do you recommend for someone who does not understand millions?*

Synthesis: *How would you design a place value chart that includes decimals?*

Rigorous Problem Solving with the Concept of Thousands and Millions 1

While playing a place value game, Samantha rolled 9 numbers and placed them in the following place value positions:

3 in the hundreds place
5 in the ten thousands place
1 in the ones place
8 in the hundred thousands place
4 in the thousands place
9 in the hundred millions place
0 in the tens place
7 in the millions place
2 in the ten millions place

Part A

Which number did Samantha create?

o 729,854,310
o 927,854,301
o 972,854,301
o 927,584,301
o 927,845,301
o 927,854,031

Part B

Using the digits 2–9 only once, create an even number that has eight digits. Place a number greater than 7 in the thousands place, a number less than 3 in the millions place, a number that is divisible by 4 in the hundred thousands place, and make sure the created number is greater than 90,000,000.

Rigorous Problem Solving with the Concept of Thousands and Millions 2

While playing a place value game, Kassidy rolled 9 numbers and placed them in the following place value positions:

4 in the hundreds place
6 in the ten thousands place
2 in the ones place
9 in the hundred thousands place
5 in the thousands place
7 in the hundred millions place
1 in the tens place
8 in the millions place
3 in the ten millions place

Part A

Which number did Kassidy create?

- o 738,965,412
- o 738,095,412
- o 738,956,412
- o 738,965,421
- o 783,965,412
- o 738,965,142

Part B

Using the digits 1–8 only once, create an even number that has eight digits. Place a number greater than 4 in the thousands place, a number less than 2 in the hundreds place, a number that is divisible by 2 in the hundred thousands place, and make sure the created number is greater than 36,000,000.

Ten Thousand Dots

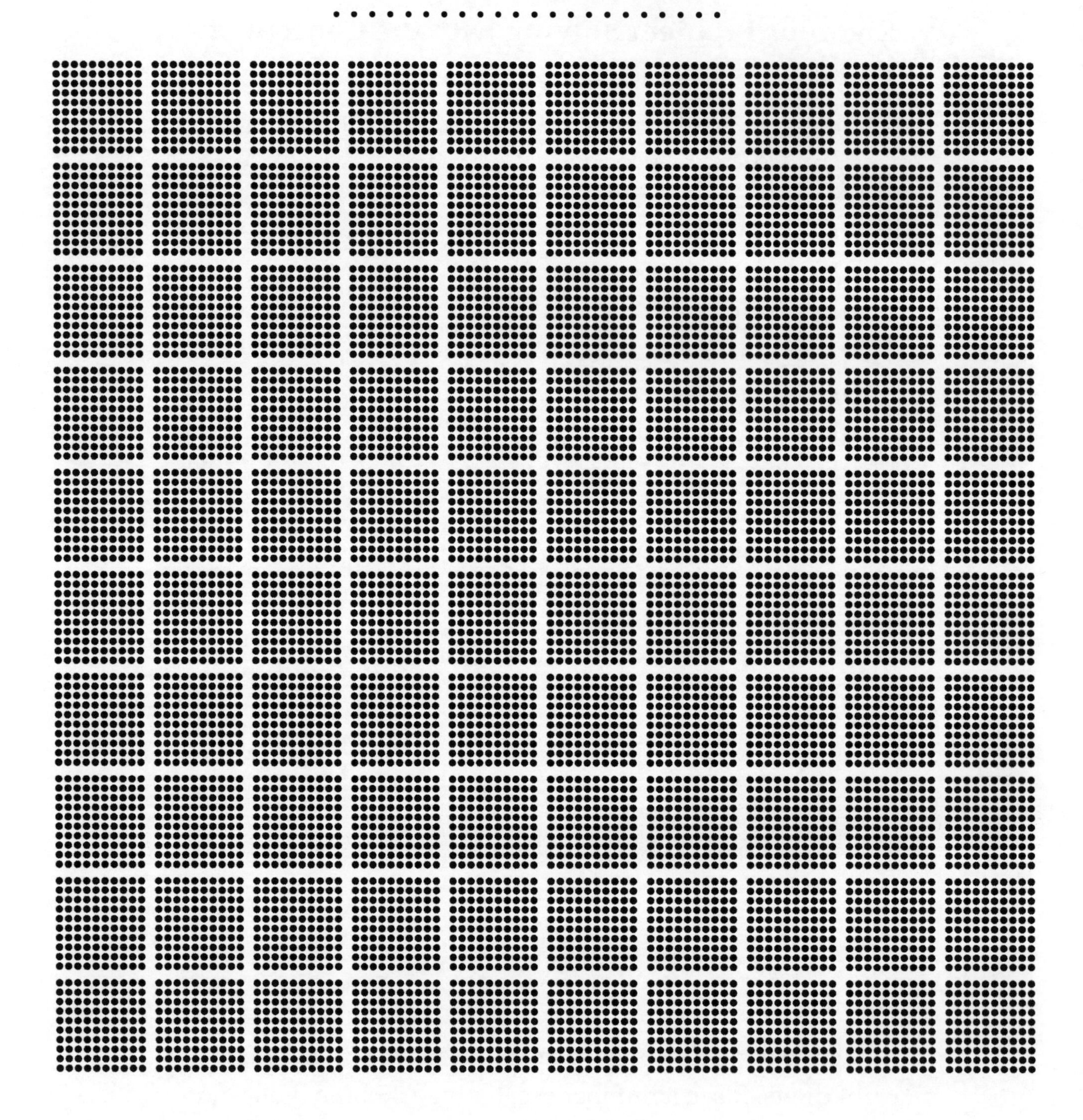

Which Place is Best?

A game designed to build the concept of thousands and millions

Materials:

- Place Value Frame for each player
- Die (labeled 1–6 or 0–9)
- Pencil or pen for each player
- File folder for each player

Directions:

1. The first player rolls the die. All players write this number in any one of the spaces on their place value frames. Players use the file folders to keep the other players from seeing where they placed the digit. Once a number is placed, it cannot be moved.
2. Players take turns rolling the die and placing digits in either the first number or the second number until all of the sections are filled.
3. Players lift the file folders to reveal their numbers. Players read and compare the numbers. The player with the largest first number earns 1,000,000 points. The player with the largest second number earns 1,000,000 points.
4. Play continues with players accumulating points until the winning player earns 10,000,000 points!

Content Differentiation:

Moving Back: Use fewer sections in the place value frames.

Moving Ahead: Include a decimal point somewhere on each place value frame.

Note: This game appears in Taylor-Cox, J. (2005) Family Math Night: Math Standards in Action. *Eye on Education.*

Place Value Frames

First number

Second number

✂- -

First number

Second number

CONCEPT:

Commutative and Associative Properties

What are the Commutative and Associative Properties?

The commutative property involves changing the order of the operations of numbers without changing the result. Students work with foundational ideas of the commutative property, also called the order property, when they work with fact families. They learn that 3 + 5 = 5 + 3 and 6 x 8 = 8 x 6. It is important for students to understand that addition and multiplication have the commutative property, while subtraction and division do not. Similarly, the associative property involves changing the grouping of the numbers without changing the result. (2 + 7) + 3 = 2 + (7 + 3) and (4 x 5) x 2 = 4 x (5 x 2). Students must understand that addition and multiplication have the associative property, while subtraction and division do not.

CCSS

Operations and Algebraic Thinking

Formative Assessment

To find out if a student understands the commutative and associative properties ask the students to identify which property (commutative or associative), if any, the following equations represent:

8 + 7 = 7 + 8
2 + (5 + 3) = (2 + 5) + 3
(5 x 2) x 3 = 5 x (2 x 3)
(5 x 2) x 7 = 7 x 3
(6 + 8) + 9 = (8 + 6) + 9

Successful Strategies

Analyzing equations to find out if the order is changed or the groupings are changed is an important aspect of understanding the commutative and associate properties. It is critical that students also experience situations that involve addition and multiplication so that they learn when the properties may not be applicable.

Math Words to Use

commutative property, order, associative property, grouping, addition, multiplication, sum, product, expression, equation

Questions at Different *Levels of Cognitive Demand*

Recall: *Does* $5 + 3 = 3 + 4$*?*

Comprehension: *Why does* $6 \times 7 = 7 \times 6$*?*

Application: *How would you show the commutative property using the digits 3, 4, 5?*

Analysis: *How would you compare the associative and commutative properties?*

Evaluation: *What is the value in knowing the commutative and associative properties?*

Synthesis: *How would you rearrange an expression that includes addition and multiplication to make a different, yet equal expression?*

COMPLEX

Rigorous Problem Solving with the Concept of Commutative and Associative Properties 1

Sandy decided to collect shiny dimes. She found 3 dimes each day for 1 week. She doubled her collection the second week.

Part A

Which equations use the associative property to represent the number of dimes Sandy found?

- o $(3 \times 1) \times 2 = 3 \times (1 \times 2)$
- o $(7 \times 3) + 2 = 7 \times (3 + 2)$
- o $(7 \times 7) + 2 = 7 \times (7 + 2)$
- o $(7 \times 3) \times 2 = 7 \times (3 \times 2)$
- o $(7 + 3) \times 2 = 7 \times (3 + 2)$
- o $(3 \times 7) \times 2 = 3 \times (7 \times 2)$

Part B

By the next month, Sandy had 100 dimes. She gave Kayla half of her dimes. She gave Kass 23 of her dimes. Does the associative property represent how many dimes Sandy has left?

$(100 \div 2) - 23 = 100 \div (2 - 23)$

Yes or no?___________

Explain your answer.

Rigorous Problem Solving with the Concept of Commutative and Associative Properties 2

David left shiny dimes for his family. He left 4 dimes each day for 2 weeks.

Part A

Which equation uses the commutative property to represent the number of dimes David left?

- o $4 \times 2 = 2 \times 4$
- o $4 \times 2 \times 2 = 2 \times 4 \times 2$
- o $4 \times 14 = 14 \times 4$
- o $4 \times 7 = 7 \times 4$

Part B

Grandpaw found 12 dimes. Don found 14 dimes. Sam and James each found 15 dimes. Does the commutative property represent how many dimes were found?

$12 + 14 + (15 \times 2) = (2 \times 15) + 14 + 12$

Yes or no? __________

Explain your answer.

Baseball

A game designed to build the concepts of commutative and associative properties

Materials:

- Baseball Spinner, paper clip
- 3 dice (labeled 1–6 with numerals)
- Baseball Field resource page
- Pawns of a different color for each player
- Dry erase board and marker for each player

Directions:

1. To begin the first round, a player spins the spinner. If it lands on single or double, the player rolls two dice. If it lands on triple or homerun, the player rolls three dice.
2. All players use the numbers rolled to write the type of equation shown on the spinner on their dry erase boards.
3. The first player to write the correct equation and call out the correct sum or product moves her pawn on the baseball field. For example, $(3 + 2) + 4 = 3 + (2 + 4)$ has a sum of nine. The player moves her pawn to third base.
4. Players take turns spinning the spinner and rolling the dice for each new round. Players can occupy the same base at the same time.
5. If a player's pawn is on second base and the player scores a single, she moves the pawn to third base. However, players can not score more than one run per round. For example, if a player's pawn is on third base and the player scores a double, she can only go to home plate.
6. Play continues until a player score five runs.

Content Differentiation:

Moving Back: Play this game using only the commutative property (singles and doubles).

Moving Ahead: Use parentheses in all equations.

Baseball Spinner

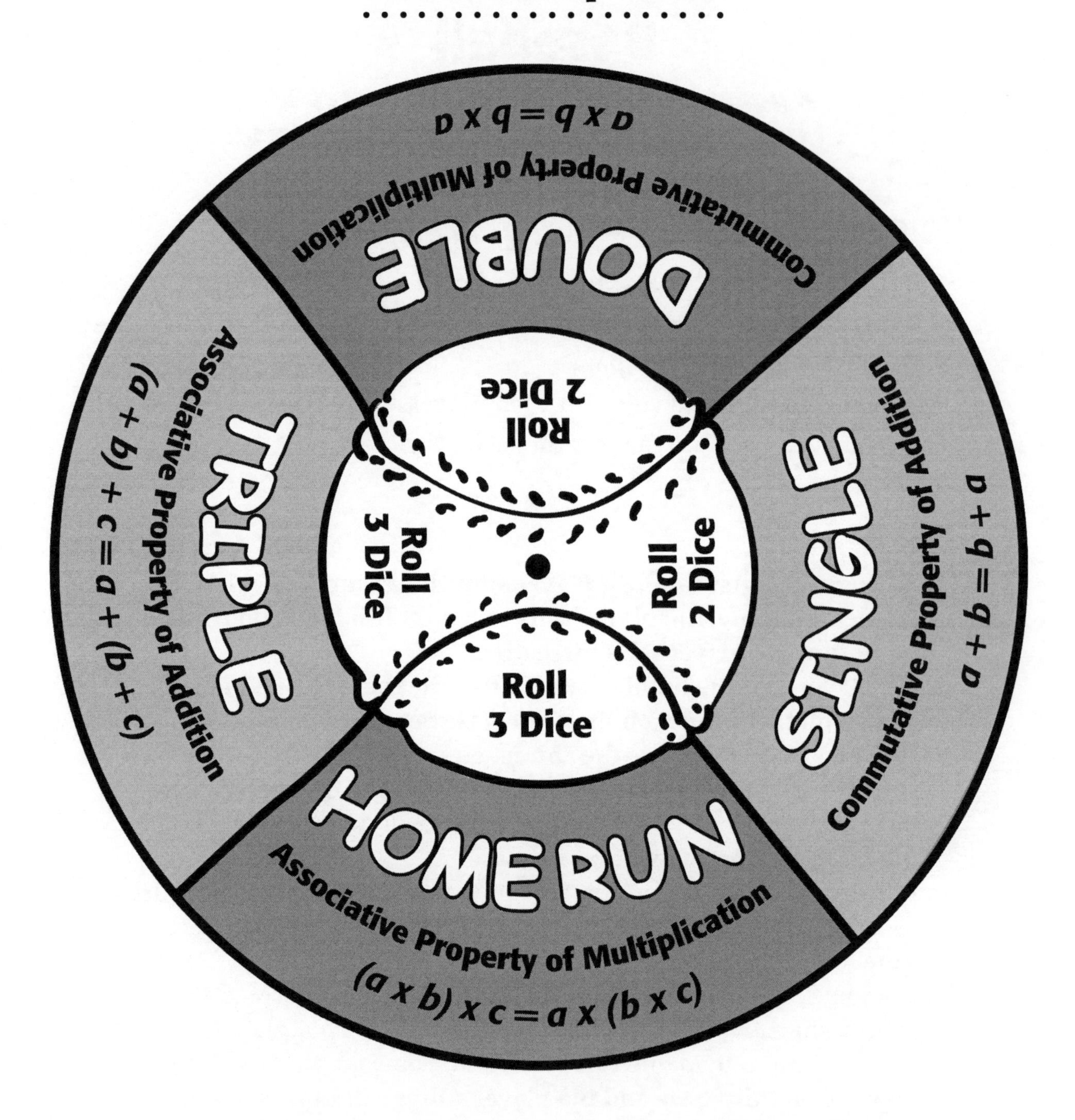

To use the spinner, place a paper clip in the center. Place pencil point through the paper clip at the center of the spinner. Holding the pencil securely with one hand, spin the paper clip with the other hand.

Baseball Field

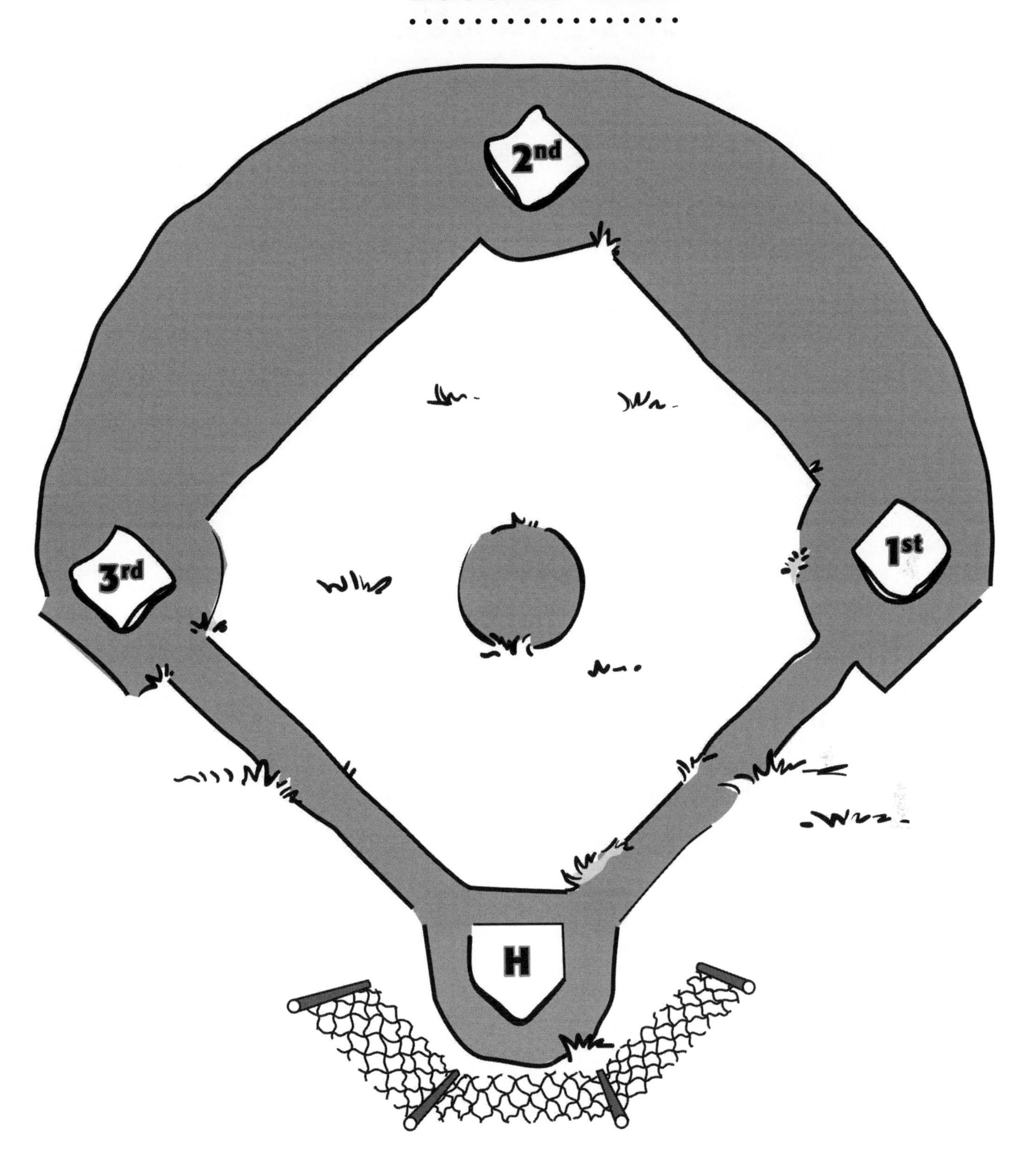

CONCEPT:
Prime and Composite Numbers

What are Prime and Composite Numbers?

Prime numbers have exactly two factors. These factors are 1 and the number itself. For example, 2 is a prime number because its only factors are 1 and 2. Composite numbers have more than two factors. For example, 12 is a composite number because it is divisible by 1, 2, 3, 4, 6, and 12. One is not considered a prime number or a composite number. Prime factorization involves renaming a composite number as a product of prime numbers.

CCSS

Operations and Algebraic Thinking

Formative Assessment

To find out if a student understands prime and composite numbers ask the students to use factor trees to show the prime factorization of the following numbers:

12 24 32 100

If the students are unfamiliar with factor trees, display the factor tree for 12 and ask them to explain the relationships they see within the factor tree as a springboard for making their own factor trees.

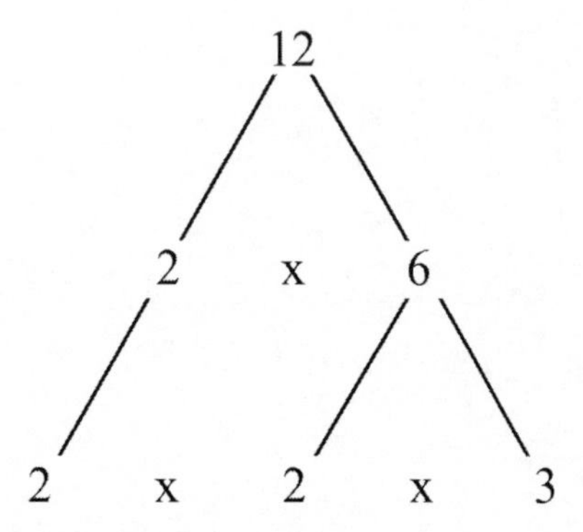

Successful Strategies

Prime factorization helps students identify and understand prime and composite numbers because students are able to see how numbers are decomposed using multiplication and division. When making a factor tree, encourage students to continue finding factors until all the factors are prime numbers. When students are ready, encourage them to record the prime factorization with exponents.

Math Words to Use

prime number, composite number, factor, factorization, factor tree, divisible

Questions at Different *Levels of Cognitive Demand*

Recall: *Is 5 a prime number?*

Comprehension: *Why is 10 a composite number?*

Application: *How would you show the prime factorization of 8?*

Analysis: *How would you compare the prime and composite numbers?*

Evaluation: *What is the value in knowing how to make a factor tree?*

Synthesis: *What if we needed to show prime factorization without a factor tree?*

COMPLEX

Rigorous Problem Solving with the Concept of Prime and Composite Numbers 1

The game *Prime Time* requires players to draw factor trees for composite numbers.

Joey drew this factor tree for the number 42.

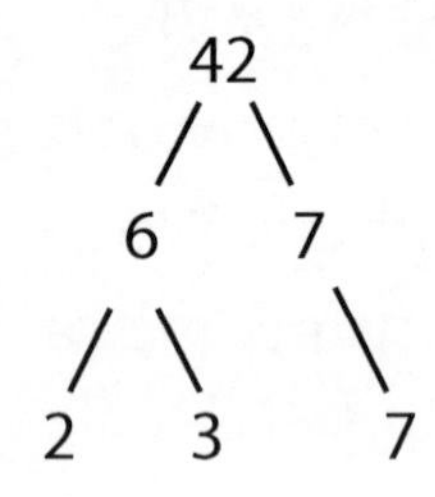

Part A

Which numbers in Joey's factor tree are composite?

- o 42, 6, 7, 2, 3, 7
- o 42, 6
- o 2, 3, 7
- o 6, 7, 2, 3

Part B

Is there another way to draw a factor tree for the number 42? If so, draw it.

Part C

If there are 42 guests and we need to arrange an even number of chairs at each table, how many different ways could we arrange the chairs?

Rigorous Problem Solving with the Concept of Prime and Composite Numbers 2

The game *Prime Time* requires players to draw factor trees for composite numbers.

Kirk drew this factor tree for the number 24.

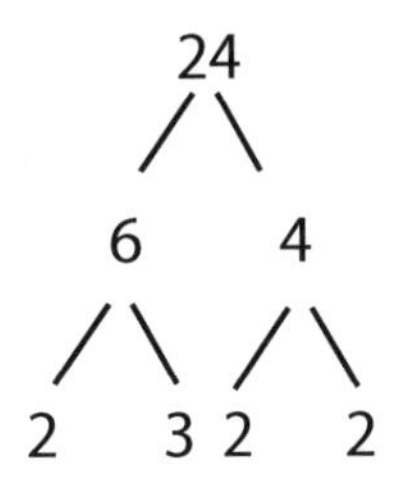

Part A

Which numbers in Kirk's factor tree are composite?

- o 6, 4, 2, 3, 2, 2
- o 2, 3
- o 2, 3, 6, 4
- o 24, 6, 4, 2, 3
- o 24, 6, 4

Part B

Is there another way to draw a factor tree for the number 24? If so, draw it.

Part C

If there are 24 guests and we need to arrange an even number of chairs at each table, how many different ways could we arrange the chairs?

Prime Time

A game designed to build the concepts of prime and composite numbers

Materials:

- Dry erase board and marker for each player
- 0–9 Number Cards copied on yellow paper *(see page 72)*
- 0–9 Number Cards copied on green paper *(see page 72)*

Directions:

1. Players shuffle each set of number cards and place them face down in two piles.
2. To begin the first round, a player chooses one yellow number card and one green number card to make a 2-digit number. The yellow card serves as the tens digit and the green card serves as the ones digit. *Note: If a yellow zero is drawn, the number will be a single digit number.*
3. If the number formed with the cards is a prime number, the player earns five points and the turn is complete. If the number formed is a composite number, the players race to draw factor trees for the number on dry erase boards. The first player to complete the factor tree calls out "Prime Factorization Complete!" If her factor tree is accurate, she earns ten points. If the factor tree is not accurate, she earns zero points.
4. Players continue taking turns choosing cards, identifying prime numbers, drawing factor trees, and keeping a running total.
5. The first player to reach a score of 50 calls out "Prime Time" and is the winner. *Note: Number cards can be shuffled and reused.*

Content Differentiation:

Moving Back: Play this game using dice (labeled 1–6) instead of 0–9 number cards.

Moving Ahead: Players write equations with exponents after they draw the factor trees.

CONCEPT:

Finding Common Multiples

What is the Finding Common Multiples Concept?

Multiples of a number are the product of multiplying that number by another whole number. For example, 18 is a multiple of 3 because 3 x 6 = 18. Three has other multiples, too: 3, 6, 9, 12, 15, 21, 24, 27, 30,. . . . Often the task is to find common multiples of two or more numbers. For example, 24 is a common multiple of 6 and 8 because 24 is a multiple of 6 and of 8. Every multiple of 24 is also a multiple of 6 and a multiple of 8. The common multiples of 6 and 8 follow an infinite pattern, 24, 48, 72, 96, 120,. . . . If the task is to find the least common multiple (LCM) of 6 and 8, the multiples of 6 and 8 are compared to find the smallest common multiple (other than zero), which is 24.

CCSS

Operations and Algebraic Thinking

Formative Assessment

To find out if a student understands the finding common multiples concept, ask the student to name at least five multiples of the following numbers:

5 7 12

Then ask the student to find at least two common multiple of the following pairs of numbers:

5, 15 4, 10 24, 16

Then ask the student to find the LCM of the following numbers:

5, 6, 15 50, 4, 20 24, 6, 15

Successful Strategies

One of the most beneficial ways to help students find the LCM is to have students show the multiples of each number in a skip counting pattern. When students examine these patterns it is easier to find the common multiples and the LCM. Students may show the multiples with numbers, words, or models. The hundred chart serves as a valuable tool in helping students to find common and least common multiples.

Math Words to Use

least, common, multiple, pattern

Questions at Different *Levels of Cognitive Demand*

Recall: *What is a multiple of 4?*

Comprehension: *Which examples could you give for the common multiples of 5 and 10?*

Application: *How would you show a pattern of common multiples?*

Analysis: *How would you compare the LCMs of two pairs of numbers?*

Evaluation: *What strategy would you recommend for finding the LCM?*

Synthesis: *When would it be necessary to find the LCM of four numbers?*

COMPLEX

Rigorous Problem Solving with the Concept of Finding Common Multiples 1

In the game *LCM Hopscotch*, players are required to find either the Least Common Multiple (LCM) or a common multiple that is not the least. Samantha drew two cards: 3 and 9.

Part A

What is the least common multiple of 3 and 9?

o 27
o 12
o 9
o 3

Part B

Write 3 common multiples of 3 and 9 that are not the least.

☐☐☐

Use equations to prove your answers.

Part C

The least common multiple of 2 numbers is 132, and the sum of the numbers is 23.

What are the numbers?

Rigorous Problem Solving with the Concept of Finding Common Multiples 2

In the game *LCM Hopscotch*, players are required to find either the Least Common Multiple (LCM) or a common multiple that is not the least. James drew two cards, 6 and 4.

Part A

What is the least common multiple of 6 and 4?

o 24
o 12
o 6
o 2

Part B

Write 3 common multiples of 6 and 4 that are not the least.

Use equations to prove your answers.

Part C

The least common multiple of 2 numbers is 77, and the sum of the numbers is 18.

What are the numbers?

LCM Hopscotch

A game designed to build the concept of finding common multiples

Materials:

- Two sets of 1–9 Numeral Cards
- Pawn for each player
- LCM Hopscotch Game Board
- LCM Hopscotch Spinner, paper clip

Directions:

1. Place pawns in the start position. Mix each set of numeral cards and place face-down in two piles.
2. The first player spins the spinner to find out if he will have a turn. For the first move, if the spinner lands on "Common multiples (not the least)" the player misses a turn. If the spinner lands on "Least common multiple" the player chooses one numeral card from each pile and flips them over. If the player names the correct LCM, he moves the pawn to the first space on the Hopscotch game board.
3. The next player takes a turn.
4. To make the second move on the Hopscotch board the player has to spin "Common multiples (not the least)." To move to this position, the player has to name two common multiples that are not the least.
5. Players continue taking turns, naming common multiples, and moving their pawns on the game board.
6. The winner is the first player to reach the top of the Hopscotch game board.

Content Differentiation:

Moving Back: Use 1–6 numeral cards.

Moving Ahead: Use three sets of numeral cards to find the LCM of three numbers.

1–9 Numeral Cards

1	2	3	4
5	6	7	8
9	1	2	3
4	5	6	7
8	9		

LCM Hopscotch Game Board

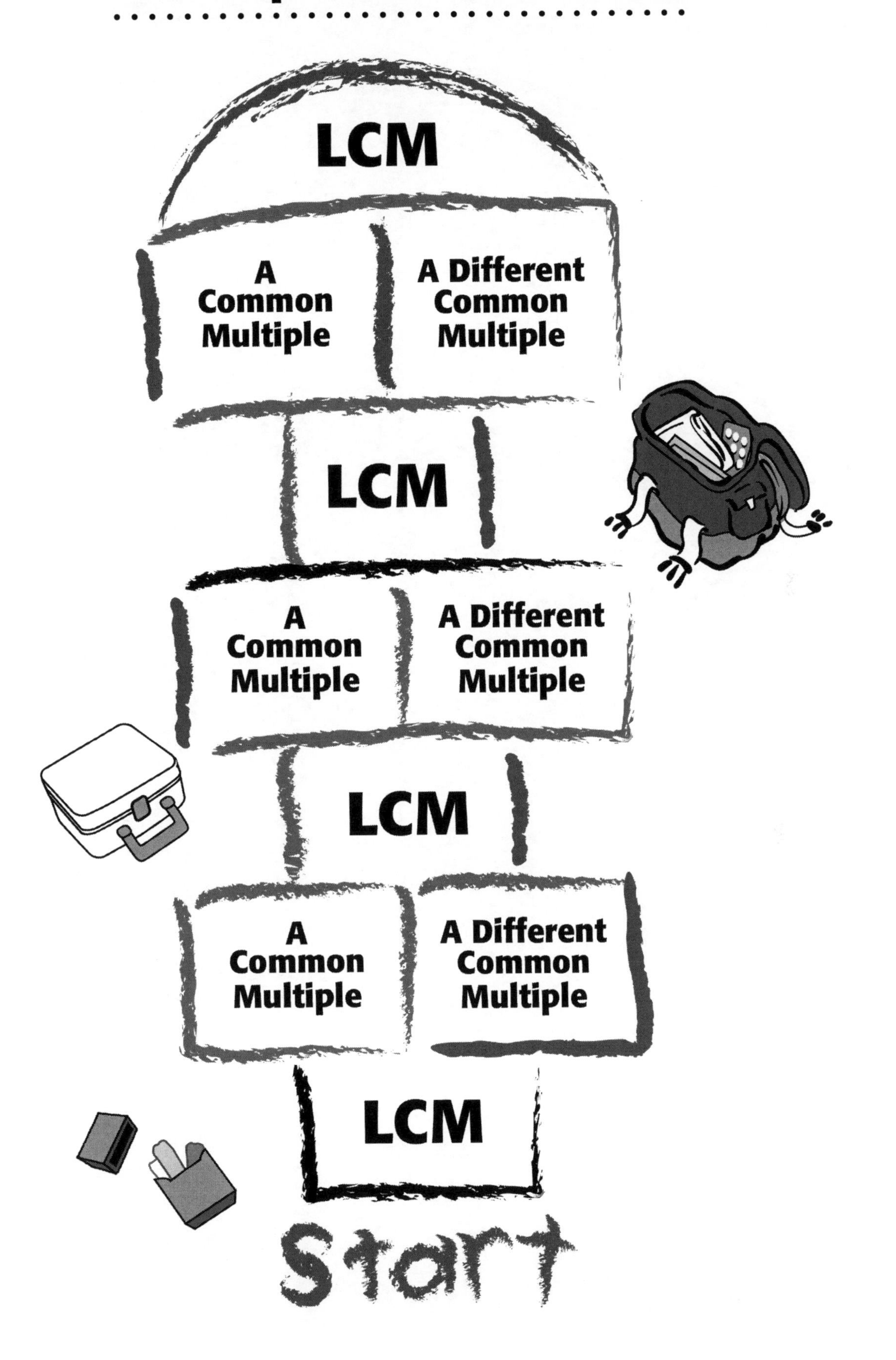

LCM Hopscotch Spinner

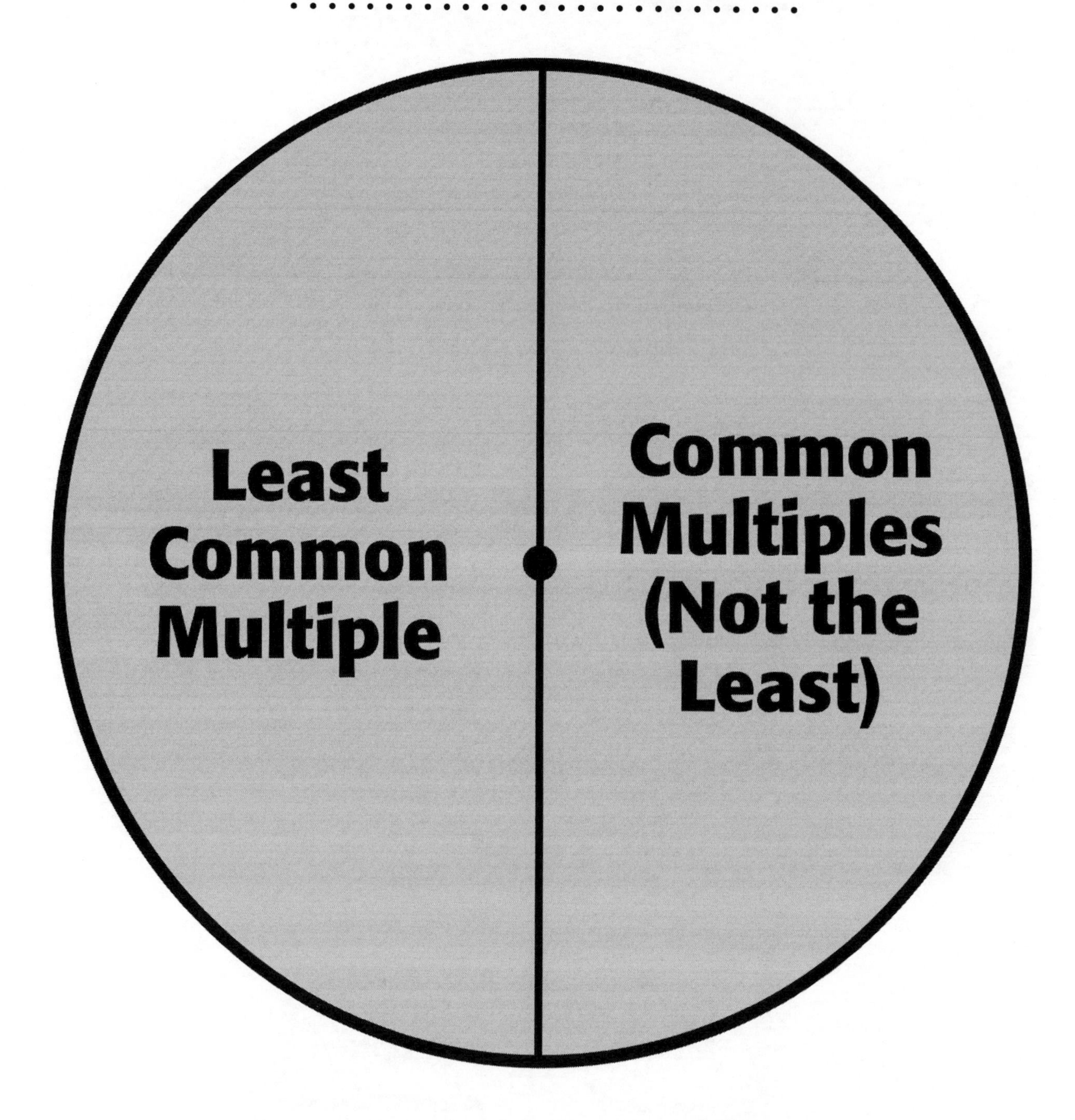

To use the spinner, place a paper clip in the center. Place pencil point through the paper clip at the center of the spinner. Holding the pencil securely with one hand, spin the paper clip with the other hand.

CONCEPT:

Finding Common Factors

What is the Finding Common Factors Concept?

A factor of a number will divide that number evenly. For example, 9 is a factor of 18 because 18 can be divided by 9 without any remainders or fractional parts of a number. Eighteen has other factors, too. It is also divided evenly by 6 and 3, also by 1 and itself. The factors of 18 are 1, 2, 3, 6, 9, 18. If a number is only divisible by 1 and itself, it is a prime number. Often the task is to find common factors of two or more numbers. For example, 5 is a common factor of 30 and 20 because 5 is a factor of 30 and of 20. In fact, 1, 2, 5, 10 are all common factors of 30 and 20. If the task is to find the greatest common factor (GCF) of 30 and 20, we compare the factors of 30 and the factors of 20 to find the largest common factor, which is 10.

CCSS

Operations and Algebraic Thinking

Formative Assessment

To find out if a student understands the finding factors concept ask the student to name at least five factors of the following numbers:

12 50 48

Then ask the student to find at least two common factors of the following pairs of numbers:

15, 35 12, 18 36, 16

Then ask the student to find the GCF of the following numbers:

9, 12, 24 18, 48, 30 33, 7, 20

Successful Strategies

As with finding the LCM, a beneficial way to help students find the GCF is to have students show the factors of each number in a skip counting pattern. When students examine these patterns it is easier to find the common factors and the GCF. Students may show the factors with numbers, words, or models. The hundred chart serves as a valuable tool in helping students to find common and greatest common factors.

Math Words to Use

greatest, common, factor, pattern

Questions at Different *Levels of Cognitive Demand*

Recall: *What is a factor of 10?*

Comprehension: *Which examples could you give for the common factors of 10 and 8?*

Application: *How would you show the order of common factors?*

Analysis: *How would you compare the GCFs for two pairs of numbers?*

Evaluation: *What strategy do you recommend for finding the GCF?*

Synthesis: *What if we needed to find the GCF of four numbers?*

COMPLEX

Rigorous Problem Solving with the Concept of Finding Common Factors 1

Captain Kirk catches crabs in the bay.

Part A

On Monday he caught 40 male crabs and 20 female crabs. Each person at the crab feast had the same number of male crabs as female crabs without any crabs being left over. How many people were at the crab feast?

List all of the possible total number of people at the crab feast.

Part B

On Tuesday, Captain Kirk went back to the bay for more crabs. He steamed 21 extra large crabs and 49 large crabs. He wanted to serve the crabs on trays so each tray will have the same number of extra large and large crabs. What is the greatest number of trays he can use to serve all the crabs?

Justify your answer.

Rigorous Problem Solving with the Concept of Finding Common Factors 2

Captain Kirk catches crabs in the bay.

Part A

On Wednesday he caught 30 male crabs and 60 female crabs. Each person at the crab feast had the same number of male crabs as female crabs without any crabs being left over. How many people were at the crab feast?

List all of the possible total number of people at the crab feast.

Part B

On Thursday, Captain Kirk went back to the bay for more crabs. He steamed 15 extra large crabs and 50 large crabs. He wanted to serve the crabs on trays so each tray will have the same number of extra large and large crabs. What is the greatest number of trays he can use to serve all the crabs?

Justify your answer.

GCF Hopscotch

A game designed to build the concept of finding common factors

Materials:

- Two sets of GCF Numeral Cards
- Pawn for each player
- GCF Hopscotch Game Board
- GCF Hopscotch Spinner, paper clip

Directions:

1. Place pawns in the start position. Mix each set of GCF numeral cards and place cards face-down in two piles.
2. The first player spins the spinner to find out if she will have a turn. For the first move, if the spinner lands on "Common factors (not the greatest)" the player misses the turn. If the spinner lands on "Greatest common factor" the player chooses one numeral card from each pile and flips both cards over. If the player names the correct GCF, she moves the pawn to the first space on the Hopscotch game board.
3. The next player takes a turn.
4. To make the second move on the Hopscotch board the player has to spin "Common factors (not the greatest)." To move to this position, the player must name two common factors that are not the greatest. If it is not possible to name two common factors that are not the greatest the player misses the turn.
5. Players continue taking turns, naming common factors, and moving their pawns on the game board. The winner is the first player to reach the top of the Hopscotch game board.

Content Differentiation:

Moving Back: Use only lower value GCF numeral cards.

Moving Ahead: Use three sets of numeral cards to find the GCF of three numbers.

GCF Numeral Cards

4	6	8	9
10	12	14	15
16	18	20	21
24	25	28	30
32	35	36	40

GCF Hopscotch Game Board

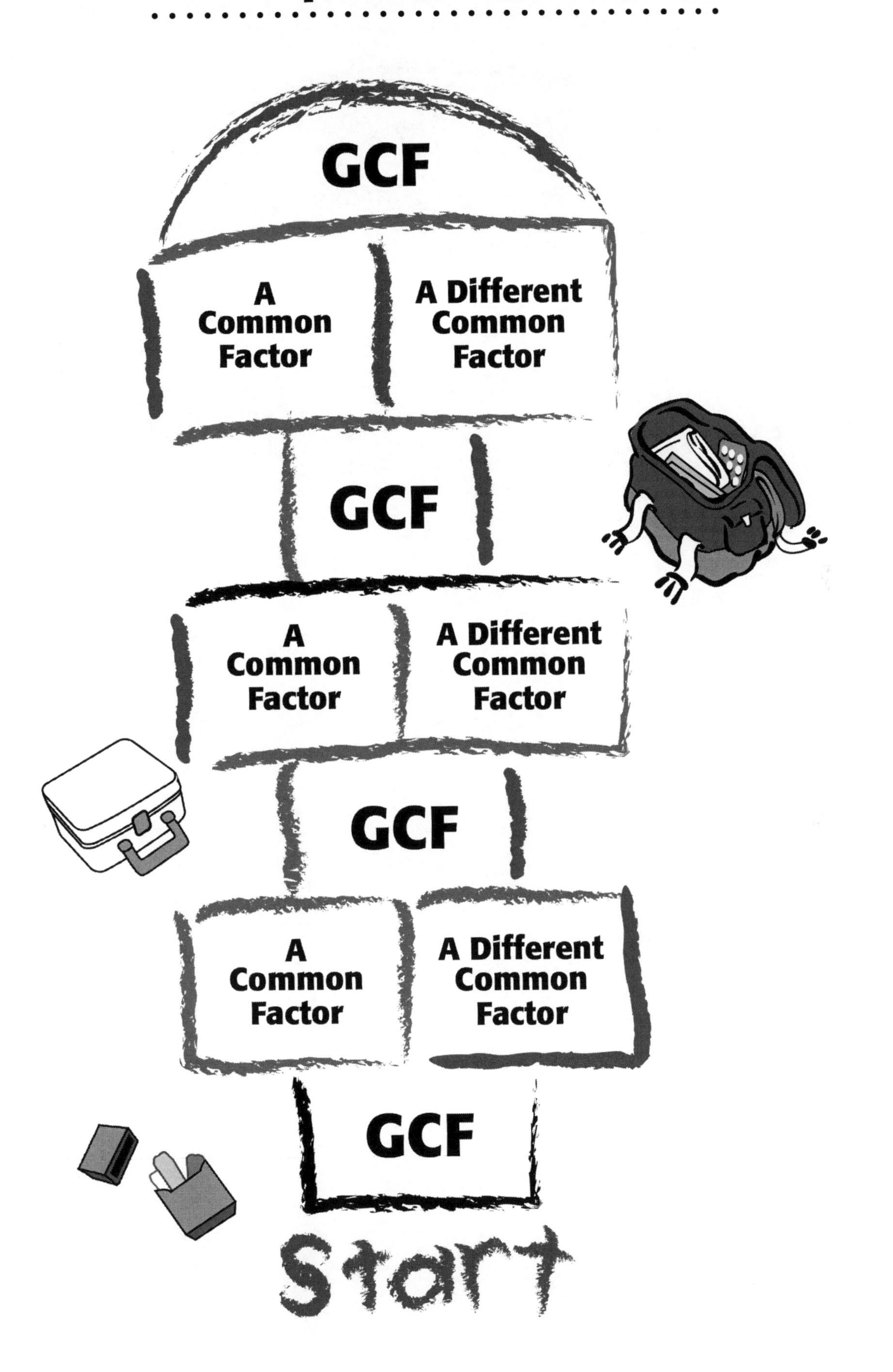

GCF Hopscotch Spinner

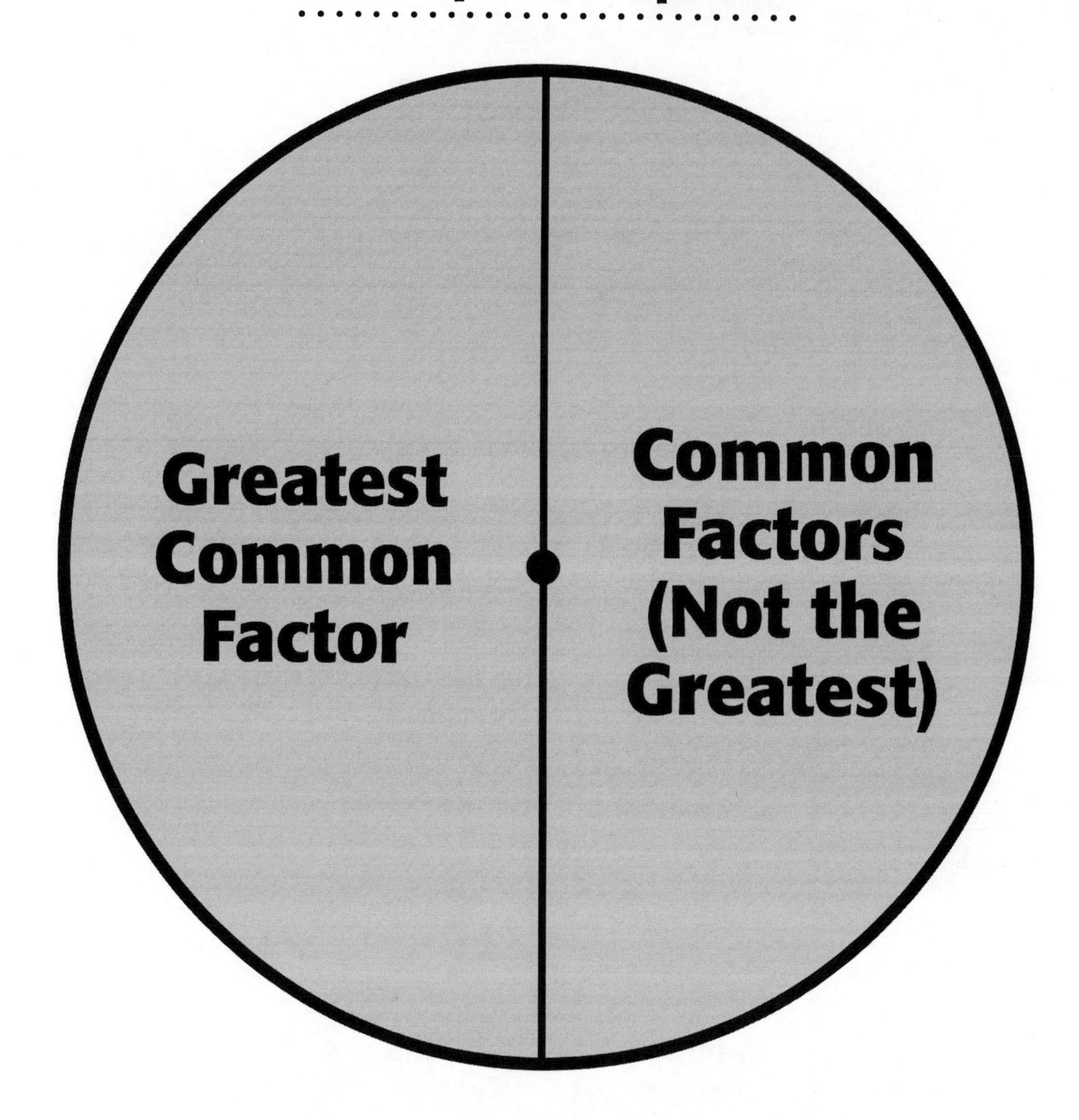

To use the spinner, place a paper clip in the center. Place pencil point through the paper clip at the center of the spinner. Holding the pencil securely with one hand, spin the paper clip with the other hand.

References

Bermejo, V. (1996). Cardinality Development and Counting. *Developmental Psychology,* vol. 32, no. 2, pp 263–268.

Bloom, B., Englehart, M. Furst, E., Hill, W., & Krathwohl, D. (1956). Taxonomy of Educational Objectives; The Classification of Educational Goals. *Handbook I: Cognitive Domain.* New York, Toronto: Longmans Green.

Clements, D. (1999). Subitizing: What is it? Why teach it? *Teaching Children Mathematics,* 5 (March): 400–405.

Freeman, F. N. (1912). Grouped Objects as a Concrete Basis for the Number Idea. *Elementary School Teacher* 8, 306–314.

Kaufman, E. L., Lord, M.W., Reese, T. W. & Volkman, J. (1949). The Discrimination of Visual Number. *American Journal of Psychology,* 62, 498–525.

McIntosh, A. J., Reyes, B. J., and Reys, R. E. (1993). A Proposed Framework for Examining Number Sense. *For the Learning of Mathematics,* 12 (3), 2–8.

National Association for the Education of Young Children and National Council of Teachers of Mathematics. (2002). *Early Childhood Mathematics: Promoting Good Beginnings, A Joint Position Statement.* Reston, VA: NCTM.

National Council of Teachers of Mathematics (NCTM). (2006). *Curriculum Focal Points for Prekindergarten Through Grade 8 Mathematics: A Quest for Coherence.* Reston, VA: NCTM.

National Council of Teachers of Mathematics (NCTM). (2000). *Principles and Standards for School Mathematics*. Reston, VA: NCTM.

National Governors Association Center for Best Practices & Council of Chief State School Officers. (2010). *Common Core State Standards for Mathematics*. Washington, DC: Authors.

National Mathematics Advisory Panel (2008). *Foundations for Success: The Final Report of the National Mathematics Advisory Panel*, U.S. Department of Education: Washington, DC.

OED2, *Oxford English Dictionary, Second Edition Online*. New York, New York: Oxford University Press.

Piaget, J. (1964). Development and Learning. In R. Ripple & V Rockcastle (Eds.). *Piaget Rediscovered*. Ithaca New York: Cornell University Press.

Piaget, J. (1970). Piaget's Theory. In P.H. Mussen (Ed.). *Carmichael's Manual of Child Psychology*, vol. 1, 3rd edition. London: John Wiley and Sons, Ins.

Taylor-Cox, J. (2001). How Many Marbles in the Jar? Estimation in the Early Grades. *Teaching Children Mathematics*, 8 (4), 208–214.

Taylor-Cox, J. (2005). *Family Math Night: Math Standards in Action*. Larchmont, NY: Eye on Education.

Taylor-Cox, J. & Kelly, B. (2008). The Doctors Disagree: Two Points of View on the Topic of "Drill and Kill" in Mathematics Education. *The Banneker Banner. The Official Journal of the Maryland Council of Teachers of Mathematics, 25 (1), 6–7.*

Webb, N. (March 28, 2002). Depth-of-knowledge levels in four content areas. Unpublished paper.

An environmentally friendly book printed and bound in England by www.printondemand-worldwide.com

This book is made entirely of sustainable materials; FSC paper for the cover and PEFC paper for the text pages.

Reprint of # - CO - 276/216/13 - CB - Lamination Gloss - Printed on 22-Nov-16 07:52